NAVIGATING
YOUR SAFETY
CULTURE
JOURNEY

MIKE D. KINNEY, CSP

Copyright © 2019

Mike D. Kinney

Performance Publishing Group

McKinney, TX

All Worldwide Rights Reserved.

All rights reserved. No part of this publication may be reproduced, stored in a retrieval system or transmitted, in any form or by any means, electronic, mechanical, recorded, photocopied, or otherwise, without the prior written permission of the copyright owner, except by a reviewer who may quote brief passages in a review.

ISBN 13: 978-1-946629-43-2

ISBN 10: 1-946629-43-X

Table of Contents

Testimonials 1

About the Cover 3

About Mike Kinney 5

Foreword 7

Introduction 9

1.0 What Is Safety Culture? 15

2.0 Leaders 57

3.0 Employees 89

4.0 Where Do I Start? 129

5.0 The Safety Culture Journey 161

6.0 Not the End … Rather the End of the Beginning 207

7.0 Final Thoughts 213

Bibliography 217

Testimonials

"Mike Kinney is right on the mark with ***Navigating Your Safety Culture Journey***. He expertly explains the importance of management commitment, thorough planning, and employee engagement as the true path to safety culture excellence. It's loaded with detailed information every company needs as it travels its safety culture journey."

~David Gibbs, Lt. Col, USAF (Retired),
Program Director, Battlespace, Inc.

"***Navigating Your Safety Culture Journey*** is a well written, highly insightful, and easy to read book that gives the readers the tools necessary to critically evaluate their company's safety culture, its implementation, and the know-how to make lasting and impactful change. The book challenges the reader to critically evaluate their safety culture and make sure it best represents the image the company wants to outwardly portray. In the end, the reader will learn that one solution doesn't fit all, and that a safety culture must be customized to the particular needs of a company. The book is a very good resource that every manager and his management team should read and have on the shelf."

~Paul Dixon, PhD., Deputy Director,
Civilian Nuclear Program Office, Los Alamos, NM.

"Mike was my Chief of Safety at a large organization doing hazardous testing. He led us on a successful safety culture journey, so you can believe what he says in ***Navigating Your Safety Culture Journey***."

~Roy D. Bridges, Jr., Major General, USAF (Retired),
NASA Space Shuttle Astronaut (Pilot for STS-51F, Spacelab 2).

Testimonials

"*Navigating Your Safety Culture Journey* is a must read for any company wanting to enhance its safety culture. I have known Mike for enough time to recognize he has the depth, experience, and ability to connect with professionals in the safety industry, from the frontline to the boardroom, and to have a powerful impact. *Navigating Your Safety Culture Journey* addresses a fairly complex topic in a user-friendly manner. Your group needs to use this great book to help improve your safety culture as well as look at ways to improve, period."

~Dr. Jeffrey Magee, CMC/CBE/PDM/CSP,
Human Capital Developer - *www.JeffreyMagee.com*.

"Mike has the ability to convey safety culture concepts that will inspire readers to engage in conversations with co-workers that drive change and make safety a personal responsibility for all. I strongly recommend that *Navigating Your Safety Culture Journey* be required reading by all members of the management team for any company."

~Stacey Alderson, CHP, Director of Health Physics,
Stoller Newport News Nuclear (SN3), Inc., Newport News, VA.

"What makes a great leader? Good leaders do many things right, but great leaders engage, connect, listen to, and include their people in the conversations surrounding the business. They stand out, push boundaries, and look to create their own solutions because no two businesses are alike. Mike's latest book, *Navigating Your Safety Culture Journey*, encourages readers to find unique solutions that can be tailored to the individual needs of each company. Don't just look for that magic bullet. Spoiler alert: there's no such thing!"

~Jeffrey Hayzlett, Primetime TV and Podcast Host,
Speaker, Author and Part-Time Cowboy – *www.hayzlett.com*.

About the Cover

Like a lot of other authors, I initially struggled with selecting an appropriate cover design for this book. After examining a series of options, I realized that the concept of a journey would be ideal, since that is one of the key thoughts throughout this book.

For as long as I can remember, I've also been fascinated with the tools utilized by ancient mariners to successfully navigate the mighty oceans of this big blue marble we call Earth. In a similar fashion, some of these same tools could have helped them navigate their own safety culture journey:

- *Sextant*: Aids with having big picture perspective.
- *Spyglass*: Provides a focused long-distance view.
- *Compass*: Assists with ensuring their efforts remain on course.
- *Dividers*: Similar to early charting of navigable waters, safety culture efforts need to be planned and measured incrementally.
- *Block and Tackle*: Some efforts may prove to be a heavy lift.
- *Rope*: When they feel at the end of the proverbial rope, they can simply tie a knot in it and hang on.
- *Map*: Enough said.

About Mike Kinney

Mike Kinney is president of Safety Culture Strategies, LLC, based in North Las Vegas, Nevada. He has spoken at numerous conferences across the United States, addressing safety culture topics, and is sought-after by companies who want to enhance their safety culture processes. He authored a series of technical papers addressing safety culture and was selected to provide an article for the highly respected *Performance 360 Magazine* that discusses the important role leaders serve in championing an effective safety culture. He is also a published author for *Dare to Be a Difference Maker*.

He is a Board of Certified Safety Professionals (BCSP) Certified Safety Professional (CSP), with over thirty years of progressive experience working with numerous organizations and companies across the United States, assisting them with development and the implementation of safety and health programs. Mike is also an Exemplar-Global Certified Lead Auditor, ISO 45001, Occupational Health and Safety Management Systems. When combined with his comprehensive background regarding integrated management systems, he is uniquely qualified to evaluate implementation of governing ISO 45001 criteria. In addition, he is a graduate of the Ziglar Institute and is a Ziglar certified Keynote Speaker, Leadership Coach, as well as a Ziglar Legacy Instructor.

His diverse experience includes design, construction, and start-up of commercial nuclear power plants, as well as private sector facility programs. Areas of expertise include: development, implementation, qualification, and validation of safety management, construction, and emergency response/recovery programs, development of hazard analysis documentation, oversight and audits of facilities and programs, operational readiness reviews, safety culture processes, accident investigation, root cause analysis, regulatory compliance, construction, worker, process, and transportation safety, maintenance, and training.

During his career, Mike has provided senior level consultative support to numerous headquarters' organizational elements within the Department of Energy (DOE), the National Nuclear Security Administration (NNSA), and the U.S. Navy. Topics included safety culture, integrated safety management, seismic safety, nuclear safety, facility safety, and worker safety. He also led a series of accident investigations as well as evaluations addressing implementation of safety culture attributes. In addition, Mike has served on Occupational Safety and Health Administration (OSHA) and DOE/NNSA Voluntary Protection Program (VPP) On-Site Review Teams tasked with determining implementation status of VPP criteria by companies pursuing this noteworthy certification.

When not traveling and assisting companies with their safety culture initiatives or serving as a speaker at conferences throughout the United States, Mike enjoys spending time with his amazing wife Mary Elizabeth. They share their home with a very large, very spoiled female Maine Coon cat named Moonshadow. He is also very active with the local chapter of American Society of Safety Professionals, as well as assisting numerous charitable organizations, including serving as emcee and/or benefit charity auctioneer for galas and fundraisers. You can connect with Mike via his company website: www.scstrat.com and learn more about the services offered by Safety Culture Strategies, LLC.

Foreword

Safety is an attitude.

At Ziglar, Inc., we have over four decades of experience when it comes to the role of attitude in the workplace. Not only does the right attitude improve performance and make life more enjoyable, in areas like safety the right attitude saves lives.

Attitude is simply the outward expression or behavior of what you believe. The right attitude and behavior produces good fruit. Think of your own safety culture. What are the fruits you and your team are producing?

> *"Attitude is a reflection of character, and character is a reflection of habit."* — Tom Ziglar

Navigating Your Safety Culture Journey is much more than a book about the right attitude. It is a book that identifies the roots you must nourish and the habits you must create in order to make the right safety culture the automatic fruit of your organization.

Mike Kinney is not only a good personal friend, he is a safety expert who understands that habits, systems, beliefs, and values all must come together to create the optimum safety culture. Knowing *what* you should do and *how* you should do it are important, but not nearly as important as *why* you should do it, as derived from your core values.

Foreword

Get ready! You are about to embark on a journey that will allow you to create the safety culture that will positively impact you and your team for generations to come.

Tom Ziglar, CEO of Ziglar, Inc.

Introduction

Throughout history, there have been varying approaches offered to enhance safety in the workplace. Unfortunately, many of the approaches focus solely on a single attribute versus examining the company processes as a whole.

Perhaps the best example of a singular approach is when a special award is provided, commonly on an annual basis, with the winner being selected from a group of workers who have not been injured. This issue can be compounded by some vendors who recommend publicizing a drawing whereby one employee can win a new vehicle if they aren't hurt during the year. One of the many downsides of this approach is that in the majority of instances, the actual number of injuries was not significantly reduced. They simply weren't reported. You'll read more about this later in this book.

There are a series of factors contributing to companies not taking a holistic approach. Examples include: lack of a strategic philosophy, not understanding the suite of factors contributing to a healthy and robust safety culture, not recognizing the need for consistent leadership and employee involvement throughout the company, and not addressing the need for long-term sustainability.

Introduction

For well over thirty years, I've had the pleasure of supporting numerous companies both in the private sector as well as in Federal Government programs and/or their contractors. During my varied career, I've worked with some amazing individuals, including researchers, presidents of organizations ranging in size from less than 100 to well over 2,000 employees, and many "free range" thinkers who routinely challenged traditional thinking when it comes to workplace safety.

While these experiences were memorable, the ones I value most are those when I had the opportunity to work directly with task level and craft level personnel. Having "carried tools" myself prior to and during the pursuit of my college education, I have gained a keen appreciation regarding the unique perspective these individuals bring to the workplace. In my humble opinion, when management is viewed as a peer by these individuals, the company has made significant progress on their safety culture journey.

Numerous individuals, including Tom Ziglar, Chief Executive Officer, Ziglar Incorporated, and Michelle Prince, Chief Executive Officer of the Prince Performance Group, challenged me regarding why I wanted to expend the significant amount of effort and resources required to author this type of book. After much reflection, my answer turned out to be fairly easy. Namely, to offer insight for companies who are interested in enhancing their safety culture. At the end of the day, if my thoughts can provide value and prove to make a difference in the workplace, it makes all of my time at the keyboard worthwhile.

Candidly, I hope you initially think that some of the topics contained in this book aren't necessarily a good fit for your company. Now, if you're like most readers, your first response to this statement is something along the lines of "What?" If I was reading this book, I'd be fairly confused right now. After all, the author should be offering the one size fits all approach.

Just buy some software, use a silver bullet solution, etc. In other words, tell me what to do!

This attitude is totally understandable because unfortunately many vendors of so-called safety solutions take that approach. From my years in the

> **Candidly, I hope you initially think that some of the topics contained in this book aren't necessarily a good fit for your company.**

safety profession, I've become keenly appreciative of the need to successfully deploy a safety culture that is tailored to the individual needs of your company. Having said that, part of the tailoring process is having the courage to consider approaches that may not feel comfortable or that you haven't heard of. The analogy I share with my clients is that I have a very large tool chest with a myriad of screwdrivers, hammers, sockets, and similar tools. Then, working as a team, we determine which "tools" are the best fit for their company. I use this approach since each company has its own unique attributes and societal norms. In turn, these attributes and norms have to be accommodated during their safety culture journey.

I also appreciate that some of my perspectives are counter to those taken by other authors (e.g., their ideas are always the best solution). However, I feel that it is important for you to be receptive to new ideas as well as work with your team to determine the overall best strategy for your company. Since the majority of companies appreciate the need to focus on the future versus solely relying on past performance, this philosophy can also assist with your efforts addressing continuous improvement. You'll also notice that I intentionally use the phrase "journey" throughout this book. This is due to recognizing the importance of long-term commitment, appreciating what will be learned throughout the process, and engaging in an ongoing effort.

Introduction

I've also included insights regarding numerous incidents, accidents, and workplace scenarios. By way of example, I was conducting a health and safety audit of a large construction site in central Arizona a few years ago. The site had approximately 50 personnel with numerous structures, including warehouses, maintenance facilities, and a Chemical Process Building (CPB). Due to the multitude of overhead piping systems that ran horizontally within the CPB, hard hats were required to be worn at all times. To enhance awareness, there were postings prominently placed on the building advising of the potential hazards and associated personal protective equipment.

While I was conducting a routine walkthrough of the CPB, I was approached by a site individual who appeared to be working in the building, asking me if I could assist him with assembling his new hard hat. I helped him with the installation of the suspension in the hard hat, making sure he also understood how to adjust the headband so it fit him comfortably. I then asked him about his duties in the CPB. His response, and my subsequent evaluation, proved very enlightening.

Similar to previous years, the site individual (Donnie) worked at the site during his summer break from college. Each year, Donnie would check with the site manager to get his work assignments, which had previously consisted of assisting with administrative duties in the site office.

However, when Donnie arrived at the site office this time, the secretary told him he would be working in the CPB for the summer. The site manager was away from the site for a series of client meetings and had forgotten about Donnie reporting to work that day. The secretary thought the site manager had already spoken with Donnie about his new work assignment.

Donnie told her he knew where the CPB was, and the secretary handed him a new hard hat, still in the plastic wrapper. Since Donnie wanted to appear that he knew what his new job was (even though he didn't), he thanked the secretary and walked across the site to the CPB. Fortunately, I encountered Donnie within a few minutes of him entering the CPB.

After speaking with the site manager, secretary, and other site personnel, I identified a series of challenges. Examples included lack of clear communications, inconsistent new hire orientations, and weaknesses regarding selection of site personnel for job assignments. Fortunately, site management was very receptive to the results of my evaluation and quickly instituted a series of process improvements. As noted previously, there are similar examples provided throughout this book. I hope they provide you with valued insight.

At the philosophical level, every company has some version of a safety culture. By definition, your current safety culture is not good or bad, per se; it is simply the one you have. Having said that, you need to determine: 1) the true implementation status of your current safety culture, and 2) if it is the safety culture you really want. Based on my years of experience and having the opportunity to work with a wide variety of companies, hopefully the information contained in this book can assist you with both of these determinations.

Due to the overarching importance of leadership to any endeavor, especially regarding safety culture, this book has been written with the company president, or equivalent, in mind. However, the leadership attributes discussed in this book can be readily applied at any management level within the organization. As such, I encourage all readers to share the information contained in this book with the entire management team.

Introduction

In a similar manner, you will undoubtedly notice different terms utilized to address personnel who perform daily work activities. Examples include workers, task level personnel, and craft personnel. In some organizations, craft personnel are referred to as bargaining unit personnel. Due to craft personnel commonly being responsible for the "hands on" work being performed, additional emphasis is placed on this group. Craft personnel are also commonly closest to the potential hazards associated with the work and truly appreciate how work is actually performed.

By the way, I look forward to hearing from you regarding how you are progressing on your safety culture journey. I am highly confident your efforts will reap many rewards, possibly the greatest of which is the enhanced relationship with your employees. So ... if you're ready ... let's begin!

You miss 100% of the shots you don't take.
— Wayne Gretzky

1.0

WHAT IS SAFETY CULTURE?

For many years, organizations did not pay a lot of attention to the overall collective health of their work environment (e.g., their culture). At the international level, the concept of safety culture gained heightened awareness after the Chernobyl Nuclear Power Plant accident/ steam explosion that occurred in April of 1986 approximately 100 miles north of Kiev, Ukraine. After that event, the following safety culture definition was developed by the International Atomic Energy Agency (IAEA): "Safety culture is that assembly of characteristics and attitudes in organizations and individuals which establishes that, as an overriding priority, nuclear plant safety issues receive the attention warranted by their significance."

Over the years, the IAEA and its subcommittees have continued to issue a series of publications addressing safety culture. Examples include: *Safety Reports Series No. 11, Developing Safety Culture in Nuclear Activities, Practical Suggestions to Assist Progress* (International Atomic Energy Agency, 1998), *INSAG-12, Basic Safety Principles for Nuclear Power Plants, 75-INSAG-3, Rev.1, A Report by the International Nuclear Safety Advisory Group* (International Atomic Energy Agency, 1999), *IAEA-TEC-*

1.0: What is Safety Culture?

DOC-1329, Safety Culture in Nuclear Installations: Guidance for Use in the Enhancement of Safety Culture (International Atomic Energy Agency, 2002), *INSAG-15, Key Practical Issues in Strengthening Safety Culture: A report by the International Nuclear Safety Advisory Group* (International Atomic Energy Agency, 2002), *Safety Reports Series No. 42, Safety Culture in the Maintenance of Nuclear Power Plants* (International Nuclear Safety Advisory Group, (2005), and *IAEA Safety Standards Series No. SF-1, Fundamental Safety Principles* (International Atomic Energy Agency, 2006).

The IAEA continues to remain in the forefront of nuclear energy, including generating more than 9,000 scientific and technical publications. Safety culture remains an area of ongoing emphasis for IAEA, including hosting an International Conference in 2016 addressing human and organizational aspects in support of assuring nuclear safety.

Within the United States, a similar emphasis regarding the need for a comprehensive safety culture arose after the investigation of the Three Mile Island (TMI) nuclear power plant accident that occurred in the spring of 1979 near Harrisburg, Pennsylvania. That investigation determined that there were a series of contributing causes, including human error. To assist with enhancing the safety of commercial nuclear power plants within the United States after the TMI accident, the commercial nuclear power industry established the Institute of Nuclear Power Operations (INPO) organization later that same year. INPO was tasked with identifying strengths and weaknesses at commercial power plants in addition to sharing best practices.

Over the next few years, INPO also examined the concept of nuclear safety culture, and it published a series of documents, including "Principles for a Strong Nuclear Safety Culture," that was issued in November of 2004. This document included a slightly different definition regarding safety culture: "An organization's values and behaviors—modeled by its leaders and internalized by its members—that serve to make nuclear safety the overriding priority."

Similar to IAEA and INPO, the United States Nuclear Regulatory Commission (NRC) also issued additional key publications addressing safety culture. Examples of INPO and NRC publications include: *NRC Regulatory Issue Summary 2005-18, Guidance for Establishing and Maintaining a Safety Conscious Work Environment* (United States Nuclear Regulatory Commission, Office of Nuclear Reactor Regulation, Office of Nuclear Material Safety and Safeguards, 2005), INPO 09-011, *Achieving Excellence in Performance Improvement, Leader and Individual Behaviors that Exemplify Problem Prevention, Detection, and Correction as a Shared Value and a Core Business Practice* (INPO, 2009), and INPO 12-012, *Traits of a Healthy Nuclear Safety Culture* (INPO, 2012).

INPO continues its ongoing mission to promote the highest level of safety and reliability in support of excellence in the operation of commercial nuclear power plants. This organization also provides training that addresses nuclear power operations, including emergency response considerations. The NRC continues to serve as the principle regulatory authority for commercial nuclear power plants within the United States in addition to medical uses of radioactive isotopes, and treatment/storage of high-level, transuranic, and low-level radioactive waste.

Collectively, these safety cultural perspectives provided the foundation for a series of improvements throughout the United States' commercial nuclear power industry as well as similar nuclear power plants located internationally (e.g., France, Germany, England). However, the ability to readily apply some of these attributes in conventional industries and routine work environments proved challenging. This was due, in part, to the initial limited distribution of the IAEA/INPO/NRC publications beyond the commercial nuclear power "family" and the majority of

1.0: What is Safety Culture?

traditional industries not recognizing the potential applicability of the concepts to their activities.

From a private sector perspective, a series of books have been authored addressing safety management and safety culture. Examples include recognizing the need for the company culture to have the capability to be resilient regarding ongoing changes, operational hazards, and associated risks (James Roughton and Nathan Crutchfield, 2014). Another book reinforces the importance of management being able to clearly articulate the need for safety excellence and associated core values (Terry L. Mathis & Shawn M. Galloway, 2013). Yet another book addresses the importance of motivation and its contribution to reduction for injury as well as appreciating the importance of having a 24/7 safety perspective that includes family members (Larry Wilson & Gary A. Higbee, 2012). Needless to say, it doesn't take long to gain an appreciation of the myriad of elements that contribute to an effective safety culture.

As a starting point, I have developed the following definition to assist readers with understanding safety culture concepts: "The safety culture of an organization is the summation of its beliefs, behaviors, and underlying assumptions." Although "safety" is included in this definition, this definition can provide a foundation for any and all efforts, including those associated with more traditional company processes. Examples include: accounting, sales, and marketing.

> ... "The safety culture of an organization is the summation of its beliefs, behaviors, and underlying assumptions."

While a balanced safety culture can be comprised of a wide spectrum of elements, the following five components have proven to provide a solid foundation for the majority of companies.

— **Flexible Culture.** The company can adjust processes as needed to address an increase in work scope or changes in work evolution/tempo, without losing focus on their key tenets.

— **Informed Culture.** Personnel throughout the company are kept abreast of information that contributes to the success of the company as well as maintaining/enhancing safety processes.

— **Reporting Culture.** Personnel recognize and embrace the need to ensure reporting of incidents, injuries, concerns, and near misses in a timely manner.

— **Just Culture.** To assist with promoting safety, personal accountability is embraced and encouraged. Individuals also understand acceptable and unacceptable behavior. Disciplinary actions taken are commensurate with the culpability of the individual involved and consistently applied.

— **Learning Culture.** The company learns from successes as well as undesired events. They can develop correct conclusions and have the commitment to revise processes to accommodate the required changes.

As noted earlier, the desired safety culture attributes can be emulated in any process within a given company, regardless of their work activities (e.g., procurement, accounting). In other words, don't take too narrow a view regarding applicability of the information provided in this book.

While there can be some variations, there are a series of key leadership attributes provided below that have proven successful for the safety culture journey of any company. I am introducing the topic of leadership early in this book in recognition of the essential role that you and your leadership team serve in this endeavor.

1.0: What is Safety Culture?

— **Leadership**. The organization adapts to the persona of the president. Therefore, it is critical that everyone understands where you want them to go.

— **Vision**. To coin an old phrase: if you don't know where you are going, how are you going to know when you get there?

— **Engagement**. An essential attribute of effective safety culture processes is the ability to engage personnel throughout the organization.

— **Communication**. The ability to have consistent and effective communication is one constant that must be part of any initiative. This is especially true when addressing the safety culture of your company.

— **Commitment**. Among the topics contained in this book is a chapter addressing speed bumps. Regardless of the initiative, or process enhancement, being addressed by your company, there will be challenges encountered. Having the ability to remain committed to the process will greatly assist with its ultimate success or lack thereof.

Although these attributes appear to be self-evident, numerous organizations have not recognized the need for them and/or have not ensured they were rigorously applied. In some instances, initiatives instituted to enhance the current safety culture have failed miserably, even with the best of intentions. That is yet another reason I chose to author this book.

While I am firmly convinced that the majority of companies can be successful, this is an ongoing journey that is not for the faint of heart and requires long-term commitment from you and your management team. Having said

> ... this is an ongoing journey that is not for the faint of heart and requires long-term commitment ...

this, I do not want to deter you from beginning the journey. However, I do need to point out that such a decision needs to be made in a deliberate and informed manner.

Believe you can and you're halfway there.

- Theodore Roosevelt

1.1

WHY IS SAFETY CULTURE IMPORTANT?

As noted earlier in this chapter, safety culture challenges have been identified as contributing to significant accidents in the nuclear power industry both in the United States and internationally. While these nuclear power plants have the potential for catastrophic accidents, there are numerous other examples, including those occurring in more "conventional" industries. A few of these instances are summarized below.

Deepwater Horizon Drilling Platform, Gulf of Mexico

In April of 2010, the Deepwater Horizon Drilling Platform, located approximately 40 miles from the Louisiana Coast, experienced a catastrophic explosion, ultimately capsizing and sinking. The drilling platform was being operated by British Petroleum (BP), cost $350,000,000 to build and $1,000,000 per day to operate. At the time of the explosion, the drilling platform was 45 days behind schedule and $58,000,000 over budget.

The accident resulted in 11 fatalities and 17 injuries, with more than 200,000,000 gallons of oil being released into the Gulf of Mexico. In

1.1: Why Is Safety Culture Important?

addition to technical challenges, there were a series of cultural issues that contributed to the accident. Examples include drilling platform personnel being hesitant to question management direction and pressure by management to continue operations. Decisions and/or explanations provided by supposedly subject matter experts also were not questioned. Ironically, similar cultural issues contributed to the Fukushima Daiichi Nuclear Power Station accident (addressed below).

Due to the magnitude of the drilling platform explosion, a series of reports were issued. Examples include: *Deepwater Horizon Accident Investigation Report* by British Petroleum and *Deep Water the Gulf Oil Disaster and the Future of Offshore Drilling* by the National Commission on the BP Deepwater Horizon Oil Spill and Offshore Drilling.

While there were a myriad of deficiencies and/or findings identified in this series of reports, programmatic challenges included lack of operational discipline, toleration of serious deviations from safe operating practices, apparent complacency towards serious process safety risks, and more emphasis placed on worker safety versus process safety. (British Petroleum, September, 2010, U.S. Chemical Safety and Hazard Investigation Board, April, 2016, National Commission on the BP Deepwater Horizon Oil Spill and Offshore Drilling, January 2011, and Deepwater Horizon Study Group, March 2011).

Fukushima Daiichi Nuclear Power Station near Ōkuma and Futaba, Japan

The Fukushima Daiichi Nuclear Power Station is located adjacent to the towns of Ōkuma and Futaba in the Fukushima Prefecture of Japan. In March of 2011, Japan was struck by a powerful tsunami centered in the Pacific Ocean, which was generated by a 9.0 magnitude earthquake.

Following the earthquake, numerous tsunamis ultimately struck the Fukushima Daiichi Nuclear Power Station.

As part of normal operations, the reactors "scrammed" (in other words, powered down), which required the plant to rely upon alternate means of power (e.g., diesel generators, backup batteries). Unfortunately, the tsunamis had destroyed the pumps that provided seawater to the diesel generators. This resulted in the majority of standby diesel generators being taken off line due to lack of seawater that served as coolant. The backup battery system was also taken out of service due to being flooded by seawater pushed onto shore from the same tsunamis. Ironically, these essential backup systems had been placed at ground level where they were subject to flooding, versus being placed on top of elevated structures, as was the practice at other plants.

While there was a minimal loss of life directly attributed to the earthquake, over 160,000 individuals had to be evacuated. Directs costs associated with the accident were initially estimated to be approximately $15-20 billion.

This accident was investigated by numerous organizations, including the Japanese government. While a series of design flaws were identified, perhaps most chilling was the acknowledgment by government representatives regarding how the "culture" was a significant contributor.

As noted by the chairman of the (Japan) Independent Investigation Commission, this event was a disaster "made in Japan." Examples included reflective obedience, reluctance to question authority, commitment to complete assign task(s), insularity, and lack of willingness to embrace outside opinions (The National Diet of Japan, the official report of The Fukushima Nuclear Accident Independent Investigation Commission, 2012). In other words, what were viewed as some of the greatest strengths of the Japanese worker could also prove to be its most significant weaknesses.

Williams Olefins Plant, Geismar, Louisiana

In June of 2013, an explosion and fire occurred at the Williams Olefins Plant located near Geismar, Louisiana. The facility produces ethylene and propylene for the petrochemical industry and commonly employs approximately 100 personnel. At the time of the accident, there were approximately 800 personnel at the plant working on a project to expand the production of ethylene. The explosion resulted in two fatalities and approximately 165 workers were injured.

The U.S. Chemical Safety and Hazard Investigation Board determined a poor process safety culture directly contributed to the accident, including lack of robust process safety management processes. The investigation team also identified weaknesses in management of change, pre-start readiness review, and hazard analysis processes (U.S. Chemical Safety and Hazard Investigation Board, 2016).

Amtrak Crash near Philadelphia, Pennsylvania

In April of 2016, an Amtrak train struck construction workers who were performing routine maintenance activities on the tracks. This accident resulted in two fatalities and approximately forty passengers being injured. The accident was investigated by members of the National Transportation Safety Board (NTSB).

The NTSB concluded that the Amtrak approach to safety placed a higher priority regarding on-time performance and created a culture of fear and bending of the rules to get the job done. In essence, Amtrak's lack of a strong safety culture was at the heart of the accident (USA Today, November 14, 2017).

Worker Injuries/Illness That Occurred During 2017

In addition to these examples that received varying levels of media attention, there are a multitude of injuries that unfortunately occur on almost a daily basis at companies throughout the United States. During 2017, the Bureau of Labor Statistics determined that there were approximately 2.8 million nonfatal workplace injuries and illnesses reported by private industry employers.

> During 2017 ... there were approximately 2.8 million nonfatal workplace injuries and illnesses ...

Due to the vast number of cases reported during 2017, there would be significant logistical challenges presented that preclude being able to readily compile a Pareto-type chart addressing those cases where lack of an effective safety culture contributed to the injury. However, it is readily acknowledged by safety practitioners, including myself, that the majority of injuries are <u>not</u> due to intentional worker error. Rather, the injuries are due to management systems that have not been developed, are ineffective, or are not being implemented as envisioned.

Unsurprisingly, workers were commonly aware of the majority of issues/problems that contributed to the illnesses/injuries but chose not to raise a concern or question the work practice. There are a myriad of influences contributing to this situation, including fear of retaliation, not wanting to call attention to themselves, peer pressure, or simply a "this is just how we do it here" attitude. In other words, lack of an effectively implemented safety culture.

Setting aside the negative impact on schedule and financial losses associated with worker injuries, the emotional impact also cannot be overlooked. I have had to respond to accidents that resulted in personnel

1.1: Why Is Safety Culture Important?

being maimed for life or not surviving their injuries. The deep psychological impacts on the family, peers, co-workers, and the management team can be long lasting.

Perhaps the most poignant event, at least for me, occurred when I was supporting a series of companies in the southwest United States. A small company and their subcontractors were responsible for construction of large storage areas (commonly the size of 2-3 football fields). The storage areas housed pea gravel and sand from prior storage locations. By definition, the work locations were very remote and many miles away from local services, including emergency response and medical. This required each location to have qualified personnel "on standby" at the site to render aide as part of their response protocols.

After construction of the new storage areas, the majority of work activities consisted of unloading pea gravel and/or sand from large belly dump trucks. The material was then transported to different portions of the storage areas through use of a large conveyor system. The conveyor belts were commonly 4-6 feet in width and could extend hundreds of feet in length, with surface speeds averaging 50 feet per minute. Due to the work environment and periodic inclement weather conditions, routine maintenance was required on all of the conveyors, including the belt tensioners. The tensioners ensured the belt surface did not slip on the rollers when loaded with the material that was being transported.

Some of the longer runs of the conveyor system required the associated belt tensioners to be adjusted "live." In other words, this adjustment had to be conducted while the conveyor belt was moving. This practice is not uncommon, and the majority of companies have established a detailed process to ensure the work can be completed safely.

In essence, a "three-man rule" is utilized whereby one individual adjusts the tensioners and another individual stands adjacent to the emergency stop switch that immediately shuts down the conveyor belt when activated. The third individual is positioned between the other two individuals with the sole responsibility of signaling in case the emergency stop switch needs to be activated. This approach ensures the individual performing the work is appropriately protected from the hazardous work conditions.

While the company had experienced a few injuries, their overall safety record was within the normal rates for this type of work. However, their safety record took a significant change for the worse during a routine maintenance cycle that was performed on a portion of the conveyor system while it was operating.

On this particular occasion, the work was to be completed during the lunch hour by a new subcontractor recently brought onto the contract. The individual performing the work was a new employee of the subcontractor and he had not performed adjustments on large conveyor systems before. Instead of applying the three-man rule addressed above, the new employee had to perform the work by himself since the majority of site personnel had left the project site to meet in town for a luncheon. This proved significant since the site personnel leaving the site to attend the luncheon included those serving in emergency response, first aid, and safety roles.

When the site personnel returned to the site after the luncheon, they found the new subcontractor employee hanging by his right ankle from the bottom side of an elevated conveyor belt that was located approximately 20 feet off the ground. The majority of bones in his body had been broken, and unfortunately, he did not survive his injuries. The injuries were so severe that his coffin had to remain closed for the subsequent

1.1: Why Is Safety Culture Important?

funeral service. Local law enforcement and medical personnel concluded that the individual had gotten snagged by the conveyor belt while it was operating, resulting in him being dragged over/under a series of belt tensioners for over 100 feet while also striking the steel framework supporting the large conveyor system.

Senior management requested that I investigate the fatality utilizing traditional accident investigation techniques that were expanded to examine safety culture considerations. Candidly, the results of this investigation were very disconcerting and uncovered a series of stressors throughout the organization. For the purpose of this book, stressors refer to work conditions that can negatively impact workers, but for which they have limited, if any, ability to correct.

One of the most significant issues identified was the heavy-handed approach taken by the site manager, who was focused on production, production, production. In this case, production translated to how many trucks were unloaded each day. This wasn't a big surprise since his personal bonus was directly linked to meeting, or exceeding, the production quota. This is a perfect example of inadvertently enabling undesired behavior through use of poor metrics. The topic of metrics is addressed in Chapter 5 of this book. The site manager also made it very clear that he did not want to hear about problems, delays, etc., that could potentially impact production.

While the site manager was aware of the required conveyor maintenance, he still directed site personnel to attend the luncheon in town since it would be a "waste of time" to have his people watch the subcontractor employee do the work. He didn't think the work was very hazardous, and therefore

> ... since it would be a "waste of time" to have his people watch the subcontractor employee do the work.

waived the requirement for the three-man rule. Due to his concerns regarding the overall site budget, the site manager also didn't want to spend the money for overtime hours that would have been required for site personnel to provide support for the three-man rule.

Site safety personnel supported this approach and the foremen told the workers to not worry about it. By the way, this is only a small sampling of the information I gained during the investigation, including interviews with the site manager. During one of these interviews, he even threatened to have me removed from the site because I was taking too long with the accident investigation and "slowing down" their ability to start operations again.

After completion of my investigation and conducting an extensive debrief with senior management, a series of actions were immediately instituted, including dismissal of the site manager. The site safety personnel and lead foremen were also replaced. The company was ultimately subject to regulatory fines as well as a lawsuit by the family of the individual who was killed.

The senior management team was understandably extremely shaken by this event, and I had a series of closed-door sessions with them while they slowly worked through their grief. The impact was felt throughout the small company. One of their independent auditors who routinely visited the site to review their programs ultimately chose to resign. This individual took the fatality very personally, feeling that he should have been able to identify some of the issues I had uncovered during his prior audits.

Although the capability to uncover such deep-seated foundational issues during performance of a traditional checklist audit is questionable at best, it further illustrates the far-reaching impacts that can result from

1.1: Why Is Safety Culture Important?

significant injuries and/or fatalities. Chapter 4 includes a discussion regarding an option to gain better insight of a given groups' strengths and/or challenges by being embedded within that group for a period of time versus relying solely on traditional audit strategies.

The company eventually resolved many of their long-standing challenges, including restructuring their bonus program. Site managers' performance (including feedback from site personnel) was routinely evaluated, and the reporting hierarchy for safety and quality personnel supporting site operations was also revised. However, these actions would not have been possible without the senior management team recognizing the need for change in their basic philosophies as well as appreciating that these types of systemic changes require long-term commitment.

I also have had to meet with presidents of companies after there was a catastrophic event at their facility. After these types of "significant emotional events," management inevitably responds with a call to action. Examples include accident investigation teams, all hands meetings, email messages, posters with renewed emphasis on safety, etc.

While these types of initiatives can provide some value after the accident has occurred, I am highly confident that most organizations would rather not have had the accident occur to begin with. In other words, a robust safety culture is in place that is being effectively implemented in a consistent manner whereby everyone is comfortable contributing to the success of the organization.

Fortunately, numerous companies will not experience a fatality or permanent disabling injury at their facilities. Although this is definitely good news, it is also important that these same companies have an appreciation of the costs associated with more "routine" injuries. Examples of these types of injuries include slips, trips, falls, strains, and sprains. In the majority of instances, these "less serious" injuries are an indication of a workplace environment that has not achieved its full potential from a safety culture perspective. While it commonly proves somewhat problematic to achieve zero injuries, companies need to appreciate that almost any injury can provide insight regarding processes that can be improved.

Ironically, these same types of injuries can rapidly escalate from relatively minor to serious, based on the resulting significance of the injury, days away from work, and similar considerations. In addition to the pain/discomfort experienced by an employee due to serious, non-fatal injuries, the costs are staggering. Based on the recently published 2018 *Liberty Mutual Workplace Safety Index*, during 2015, companies experienced losses in excess of nearly $60 billion in direct workers compensation cost (Liberty Mutual Group, 2018). In other words, more than $1,000,000,000 per week. You are reading this statement correctly, over one billion dollars per week due to serious non-fatal workplace injuries. This figure does not include financial impacts due to loss of productivity and negative impact on morale.

While these figures are staggering by any yardstick measure, they pale in comparison when compared to the latest projections regarding costs associated with the Fukushima Daiichi Nuclear Power Station addressed earlier in this chapter. Based on a Reuters news article from December 2016, those costs have more than quadrupled, with a new estimate of $188 billion (Reuters World News, 2016). However, the Japanese Minister of Economy, Trade and Industry (METI) cautioned that the costs may go even higher due to new developments and unforeseen considerations.

Hopefully, this series of examples provides a strong case regarding the need for an effective safety culture. I am highly confident that many leaders reading this book have to pause at this point and catch their breath, attempting to process some of the staggering financial impacts being realized. Speaking of leaders, since they are instrumental in facilitating institutional change within their company, the topic of leadership is addressed from a series of perspectives throughout this book.

Opportunity is missed by most people because it is dressed in overalls and looks like work.

– Thomas Edison

1.2

THE NEED TO ENSURE A "JUST" SAFETY CULTURE

To further illustrate the many facets of an effective safety culture, the concept of a "just" culture also needs to be addressed. Having a "just" culture (e.g., fair, equitable) is critical to ensuring employee performance and accountability is consistently addressed. It refers to a collective mindset whereby a questioning attitude is promoted, people are committed to excellence, complacency is avoided, and personnel accountability is embraced.

This philosophy also recognizes the need for personnel to be encouraged to promote safety, but these same individuals also understand acceptable and unacceptable behavior. In addition, the potential for honest errors is acknowledged and dealt with differently than intentional acts. These concepts have been addressed by a series of recognized leaders, including James Reason, PhD, who is one of my personal favorites. For over 25 years, Dr. Reason has focused his research on human error as well as how organizational processes contribute to events and accidents (James Reason, 2013, 2016). Industries addressed during his research

1.2: The Need to Ensure a "Just" Safety Culture

have included aviation, commercial nuclear power, marine operations, and chemical processing plants.

Dr. Reason has also authored a series of publications addressing human error and causal effects. An intriguing aspect of his research is the development of a series of tiered questions. These questions, referred by many as Reason's Unsafe Acts Algorithm, assists with determining systems challenges within the company. These same questions can also assist with differentiating between inadvertent and intentional acts by the individual. This topic is further addressed later in this chapter.

The Just Culture mindset ensures the organization learns and improves by identifying and examining its own weaknesses in a candid manner. Personnel are advised of responses that have been/will be taken to address identified weaknesses. In a Just Culture, personnel feel safe from reprisal when voicing concerns about safety or proposed work activities. These individuals also are encouraged to discuss their actions, or the actions of others, that may have contributed to an incident or accident.

Another key aspect of a Just Culture is viewing human error as a symptom of an organizational challenge versus simply blaming the individual. Past practices would examine human error from the perspective of focusing on what the individual did incorrectly, why they didn't

> ... a Just Culture is viewing human error as a symptom of an organizational challenge versus simply blaming the individual.

follow procedures, etc. Based upon the insight provided by numerous authors, including Mr. Sidney Dekker, there is a need for human error to be viewed as an indicator of systemic issues, including the operational environment and societal contributors (Sidney Dekker, 2002).

It is critical that management also ensures that willful and/or reckless actions are dealt with differently versus those human errors that were "induced" due to environmental/situational conditions beyond the direct control/influence of the individual.

Induced environmental conditions can include procedures that are not accurate, cannot be implemented as written, or are outdated. Induced situational conditions can include schedule pressure, having to compensate for personnel who are not at work, not being familiar with the tasks to be performed, etc. The main consideration is to recognize there are numerous factors that contributed to the occurrence of an undesirable event (e.g., incidents, injuries).

By way of example, there were a series of contributors to the Three Mile Island event discussed earlier, one of which addressed how switches and instruments were placed on the surface of the control panels. In some instances, key gauges could not be readily observed by the operators because they were blocked by other instruments. Other gauges did not include a visual annunciator (e.g., warning light, flashing beacon) that could have assisted with enhancing operator awareness of potential adverse system conditions. While some systems did have audio alarms, the tonal quality between those alarms was very similar. This made it difficult for operators to be able to readily differentiate between alarms that could require a different response by the operator.

One approach that has raised awareness of man/machine interface is the emergence of ergonomic considerations. While repetitive motion is one area of focus, ergonomics can also assist with configuration of workstations to reduce the potential for an ergonomic injury or operator error to occur.

1.2: The Need to Ensure a "Just" Safety Culture

The concept of a Just Culture has had an ongoing emphasis within the commercial aviation industry for many years. An organization that has been instrumental with this effort is the Global Aviation Information Network (GAIN). This group was established to help the aviation industry work in a collaborative manner with government and labor representatives to enhance safety programs. GAIN has published a series of documents, including one entitled *A Roadmap to a Just Culture: Enhancing the Safety Environment* (Global Aviation Information Network, 2004). This publication expanded upon the algorithm concept established by Dr. Reason, who also served as an independent reviewer.

This philosophy was reinforced in a sister publication (Flight Safety Digest), via an article also entitled *A Roadmap to a Just Culture: Enhancing the Safety Environment*. It reinforced the importance of learning from unsafe acts, clearly articulating unsafe behaviors and methodologies to consistently determined negligence. This publication also included insight from case studies conducted in New Zealand, the United Kingdom, the United States, and Denmark (Flight Safety Foundation, 2005).

As noted earlier, to be truly effective, a Just Culture needs to clearly define acceptable as well as unacceptable behavior (e.g., unsafe acts). While this may appear counterintuitive (after all, everyone should know what unacceptable behavior is, right?), having clear behavioral criteria supports being able to consistently evaluate accidents/incidents for disciplinary action considerations. The following types of unsafe acts are commonly addressed as part of the evaluation process:

— **Inadvertent human error**: This type of unsafe act is determined to be unintended conduct by the employee. Disciplinary actions are commonly minimal.

— **Negligent conduct**: The employee did not utilize the reasonable level of skill that would be considered the "norm" within their group, company, or industry. Disciplinary actions may include counseling, additional training, or issuance of a written warning.

— **Reckless conduct**: The unsafe act performed by the employee was not justified and the risks from such an act would be readily obvious to a reasonable person. Some companies define this type of unsafe act as gross negligence. Disciplinary actions may include time off without pay, demotion, or reduction in salary.

— **Intentional violation**: Simply put, the employee understood the results of their unsafe act, but chose to continue with their actions. In other words, the unsafe act was willful. Due to the level of egregiousness associated with this type of unsafe act, immediate suspension and/or termination is commonly considered.

Ideally, the concepts addressed above are also woven into the overall human resource processes within the company. Maximum value is gained when task level employees are afforded the opportunity to participate in development of the disciplinary actions criteria, or at a minimum, are fully apprised of the consequences for performing unsafe acts. By the way, use of these concepts addressing unsafe acts can assist your company with determining the current status of your safety culture mindset as well as next steps to be taken that can enhance same in outlying years.

Without a Just Culture mindset, personnel can very quickly realize that management is not willing to examine the systems governing the requisite work activity. While there is a significant level of effort associated with taking a systems approach, the end result can far outweigh the investment.

1.2: The Need to Ensure a "Just" Safety Culture

Conversely, while there is significantly less effort required to simply blame the worker, the results can prove catastrophic, including personnel choosing to not report injuries or near misses. Ultimately, the majority of safety culture initiatives will falter or, at a minimum, not realize their full potential, unless the concept of a Just Culture are part of the process.

Whether you think you can or think you can't, you're right.

— Henry Ford

1.3

Examples of Safety Culture Implementation Levels

Similar to the leadership styles addressed in Chapter 2, there are also varieties of safety culture implementation levels that can be encountered. A series of organizations and authors have provided a myriad of examples, which are summarized below:

— **Emerging**: Management has begun to understand the importance of management commitment regarding development of an effective safety culture.

— **Managing and Involving**: Management appreciates the importance of personnel being involved in safety culture processes.

— **Co-Operating**: Management serves as a champion regarding company personnel being committed to safety.

— **Continually Improving**: Management places emphasis addressing continuous improvement and avoiding complacency.

1.3: Examples of Safety Culture Implementation Levels

While these generic examples can provide some limited value, a series of industries have placed a focused effort on this topic, including the hospital, care giving, and aviation industries. One of my personal favorites is the safety culture implementation levels identified by the National Aerospace Laboratory NLR Air Transport Safety Institute, located in Amsterdam, The Netherlands. The NLR Air Transport Safety Institute also recognized the importance of a Just Culture, including application of those culture concepts for ground service personnel working at airports. Tasks for ground service personnel include handling of luggage, baggage, cargo, and similar ramp activities.

In concert with efforts of the GAIN group addressed in Chapter 1.2, the NLR Air Transport Safety Institute has published a series of documents, including NLR-TR-2010-431, *Just Culture and Human Factors Training in Ground Service Providers* (National Aerospace Laboratory, NLR Air Transport Safety Institute, 2011). This publication further expanded application of the human factors concepts developed by James Reason by establishing a series of safety culture implementation levels, which are summarized below.

Level 1, Pathological: At this level, safety is discounted regarding its importance. Safety is not part of project planning and is only considered after a serious injury has occurred. Company level communications regarding safety are only developed after a serious event and are short lived at best. Workers are commonly viewed as the cause for any and all accidents, organizational contributors are not considered. Short cuts taken which compromise safety, and/or not following procedures to meet mission, routinely occur and are rewarded. Individual suggestions regarding potential process improvements and/or safety concerns are not emphasized, recognized, or supported. There is a lack of interest or willingness to proactively address known safety hazards or risks.

Hopefully, you won't encounter this implementation level within your organization. However, there are unfortunately a series of companies that choose to not embrace ownership of safety programs at the management level. Due to the regulatory fines and/or litigation that such a company can be subject to, this approach commonly does not survive the test of time.

Level 2, Reactive: At Level 2, safety is commonly viewed as a necessary burden that is being imposed on the organization. Minimal compliance is the norm. In other words, only the absolute minimum is done to meet regulatory requirements versus considering additional requirements that could provide value to the organization. Actions to address known safety hazards or risks are only initiated after a serious event has occurred. Similar to a company with Level 1 safety culture implementation, unsafe behaviors, including non-compliance with procedures, can be endorsed or even rewarded.

Some companies attempt to justify this implementation strategy due to lack of budget, not having sufficient safety personnel available, etc. It is also not uncommon for the safety department to be held accountable for injuries that occur in the workplace, even though this group is not responsible for performance of day-to-day work activities. While not ideal, at least this type of company is not at Level 1 Implementation.

1.3: Examples of Safety Culture Implementation Levels

Level 3, Calculative: At this level, safety has begun to become a factor that has to be addressed, including during project planning processes. While safety has started to be considered during decision-making, safety itself is not recognized as a core value. Investigation of events and/or injuries has a component that examines organizational processes that may have been contributors. While a safety reporting process is in place that meets regulatory requirements, it is primarily utilized to gather information after an event/injury. Operational risks that are perceived as more significant are being addressed. While there are occasional deviations from following safe work practices, this is not the norm.

The status of safety culture implementation for numerous companies is commonly at this level. Many companies with Level 3 implementation also struggle with consistently identifying risks regarding significance. This can be due to a series of challenges, including lack of a comprehensive risk management identification system that addresses all potential operational risks. In turn, they do not have a consistent method to recognize which risks needs to be dispositioned in the near term.

Level 4, Proactive: At Level 4, the concept of safety is viewed as a prerequisite for organizational activities. Safety is a highly visible core value that plays a key role in the decision-making process from the highest level of the organization to those at the task level. The safety reporting process includes items that are minor in nature, and actions taken to address reporting are monitored by management. After an event occurs, there is emphasis on reducing the potential for a similar event to be realized in the future. Deviations from safe work practices are not routinely observed nor tolerated. There is an enhanced sense of operational risks, including actions that are needed to mitigate them.

Over the last few years, a series of companies have begun to achieve this level of safety culture implementation. Company processes supporting this level of implementation can also provide value when pursuing external certifications. Examples include the DOE/OSHA Voluntary Protection Program (VPP), and those offered by the International Organization for Standardization (ISO). ISO certifications include ISO 9001, Quality Management Systems; ISO 14001, Environmental Management Systems; and ISO 45001, Occupational Health and Safety Management Systems.

Level 5, Generative: When a company has achieved a Level 5 safety culture, safety has expanded beyond just a core value to being recognized as critical to continuity of operations. Acceptable and unacceptable behaviors have been established and are modeled throughout the company. Actions taken to address events/injuries recognize the concept of culpability by the worker as well as contributing factors that are beyond control of the individual. There is a high level of trust within the company, including active use of the safety reporting process. Safety concepts remain on the forefront throughout the organization, including actions to be taken to mitigate unacceptable risks. Deviation from safety processes is not tolerated. Personnel responsible for performance of the work appreciate the potential for an accident to occur and maintain a high level of situational awareness.

1.3: Examples of Safety Culture Implementation Levels

Obviously, this is the ideal state for any company. However, I advise caution attempting to establish this level of performance as your desired end state at the beginning of your safety culture journey. It is important to not over reach (e.g., having a bridge too far) to prevent your team from becoming frustrated or burned out when success is not quickly realized. Therefore, companies with Level 2 or 3 implementation are encouraged to initially pursue the next higher level of implementation. In other words, Level 2 companies should consider pursuing Level 3 implementation prior to pursuing Level 4 implementation.

As the leader for your company, you are encouraged to include these implementation levels definitions, edited as needed to reflect your company tenets, during your safety culture journey. By way of example, I have utilized these implementation levels during discussions with the management team for a series of companies. One of the questions I have found to be most insightful is when the management team is asked why they feel the company is <u>not</u> at Level 1 or Level 2 implementation. In addition to reinforcing the importance of not having major program deficiencies, this approach can assist with identifying key safety culture strategies that are currently serving as strengths for the company.

Another approach that can prove beneficial is to include these implementation levels during initial brainstorming sessions with your management team. Prior to your first team meeting, you are encouraged to include these implementation levels as part of pre-planning information routed to attendees. Valued insight can be gained by asking your management team to make their own selections regarding current implementation levels. In turn, this information can be summarized for use during your initial team meeting and beyond.

These implementation levels can also serve as a roadmap to assist with near-term activities and long-term continuous improvement initiatives.

Binning of company processes with the desired implementation level can further reinforce the importance of the collective efforts.

I have also found that these implementation levels can provide valued insight as your company continues its safety culture efforts. Examples include assisting with revising your desired end state as your safety culture matures and developing interim goals. As such, this sub-chapter is periodically referenced throughout this book. As a gentle reminder, the most value is gained from what is learned during your safety culture journey. Therefore, you need to avoid pressuring your team to "hurry up and finish." Refer to Chapter 4.3 for additional insight regarding practicing deliberate speed.

It does not matter how slowly you go as long as you do not stop.

– Confucius

1.4

How Do I Know if I Have a Good Safety Culture?

Great question! As discussed earlier, every company has some version of a safety culture. Their safety culture is necessarily not good or bad per se. It is simply the safety culture that currently exists. However, is it the safety culture they really want?

The willingness to be quizzical regarding what an "effective" safety culture looks like is essential to recognizing what is entailed to help guide your journey. A good safety culture is kind of like art, you know it when you "see" it. The main difference is that the desired attributes of a good safety culture are predominately not nearly as visible as a piece of framed art hanging on the wall.

> A good safety culture is kind of like art, you know it when you "see" it.

This conundrum is compounded by numerous safety program vendors/consultants offering the "perfect" solution to solve your cultural issues. By the way, additional confusion can be introduced since

1.4: How Do I Know if I Have a Good Safety Culture?

utilization of generic surveys, checklists, etc., may result in some temporary improvement of the safety culture. However, these generic quick fixes, by definition, do not provide for company specific considerations. As a result, they commonly do not result in long-term success.

So … if these off the shelf, one size fits all, approaches do not provide long-term value, what is the ideal solution? At the risk of sounding like an attorney asked to review your tax return, it really depends. Having said that, the following are a few big picture questions that need to be considered:

— **What is the current status of the relationship between management and the employees responsible for performance of the work?** Here's a hint: if you cannot definitively answer this question with confidence, you have already identified a key step in your journey.

— **How is performance recognized?** The majority of companies have methods in place to recognize performance. Awards may be monetary in nature, gift certificates, company apparel, and similar techniques.

One of the companies that owns a series of casino properties in Las Vegas, Nevada, has an annual Employee of the Year Award, with the winner receiving $20,000. The nominations, which are submitted by their peers, must address how the nominee demonstrated meeting the company's vision and goals regarding customer service. While this level of financial award is noteworthy, the company also ensures that the four runners up receive a monetary award. The program also includes monthly and quarterly awards. Here's another hint: when examining other performance award options for craft personnel (e.g., construction workers, welders, electricians) the two most popular items are food and apparel.

— **How are disciplinary actions addressed?** As discussed earlier, the need for a Just Culture also needs to come into play when examining disciplinary processes. Unfortunately, this topic has proven to be a significant challenge for many companies.

By way of example, one of the companies I worked with had a "mandatory" disciplinary action program that was to be followed if personnel were cited by the police for an at-fault accident while operating a company motor vehicle. At a minimum, without exception, the driver was to be given at least one day off without pay. In other words, the individual could not use vacation hours or sick time for the day off. However, on numerous occasions, the driver's line manager (e.g., project manager, supervisor) would decide not to follow the "mandatory" disciplinary action. A series of justifications were provided, including the driver didn't mean to cause the accident, s/he really is a good driver, s/he had not had an accident before, or s/he really needs her/his paycheck.

Most disturbing is that senior management was aware of this practice and did not intervene, deciding the decision needed to be rendered at the lower level. Senior management also did not want to indicate that they were overriding the decisions of their direct reports. While empowering middle management is commendable, this strategy resulted in significant frustration and discontent at the task level, since people having the same type of motor vehicle accident were not disciplined in the same way. Here's another hint: anytime senior management actions impact the paychecks of their employees, the need for consistent and equable actions (e.g., just) are even more paramount.

— **Do you know what your employees really think?** Unsurprisingly, the majority of company level managers are confident that they "know" the answer to this question. However, the actual feeling at the shop floor level can be diametrically opposed on numerous occasions. In some instances, the perceptions of the shop floor personnel are not based on actual facts and may not reflect reality. However, senior

1.4: How Do I Know if I Have a Good Safety Culture?

management needs to realize that this perception is "real" to those employees and therefore must be addressed in an open and candid manner.

Throughout this book, thoughts are provided regarding safety culture examples, from both a positive and negative perspective. Here are a few to get started, so to speak.

○ **Casino property, Las Vegas, Nevada:** I have had the pleasure to speak at numerous safety conferences and similar events and am a zealot regarding my appearance, including my footwear. As such, I commonly seek out a shoeshine shop to ensure my dress shoes are freshly polished. After leaving my vehicle with the valet service at the casino property where the conference was being held, I immediately searched for a shoeshine service, which was located fairly close to the casino's convention center.

Upon arrival at the shoeshine service, I was greeted by a very friendly young man of Russian decent. After having me climb into the chair, the young man began to clean my shoes prior to applying new polish. He then became very concerned after looking at my shoes and asked me if he could take the time to remove all of the old polish since it had not been applied properly. After I agreed, he then proceeded with his task. The final result was absolutely fantastic, and I made sure to compliment him on his efforts.

Since he essentially had to shine my shoes twice, I offered to pay more than his usual rate. The young man graciously declined, stating he just wanted to make sure I had a good shine on my shoes. Then he got very serious and inquired if I had gotten my shoes shined at his property (his words, he was very proud to work there) the last time. When I responded that I didn't, he was very relieved, saying, "Good, otherwise I was going to tell my boss that someone did not do their job right." One could say he really "owned" his work.

- **Major hotel, Dallas, Texas:** I was attending a national safety conference where I had been asked to serve as speaker for a series of sessions addressing safety culture. Due to having a fair amount of luggage, I was provided with a bellman (George) to assist with getting the luggage to my room.

While waiting for the elevator (seems like I always manage to check in when all of the elevators are busy), I struck up a quick discussion with George. He was well aware of the large safety conference being held at the hotel and was curious regarding the safety topics I would be speaking about during the week. I shared some of my thoughts on safety culture and management commitment, and then inquired how long he had worked at the hotel, what he liked about his job, etc.

He was very enthusiastic when discussing how much he enjoyed meeting the guests and helping them enjoy the best experience possible during their stay. He even joked about not having to go to the local gym since he had to lift heavy bags most of the day. He had worked at the hotel for approximately five years and appeared very interested in being promoted when the opportunity arose.

George was also very excited when talking about his "former" boss. He shared a series of examples, explaining that the boss would spend a lot of time under the porte-cochere, thanking the bellmen for their efforts, and calling out to them by their first names. He even pitched in to help greet arriving guests and unload their luggage during times when the hotel was extremely busy. He then told me that his former boss had been promoted approximately six months prior, which included being relocated to another hotel property.

1.4: How Do I Know if I Have a Good Safety Culture?

The most interesting part of the conversation for me was when I asked him about his current boss. His body language completely changed. He released a small sigh, and his shoulders slightly drooped. When I inquired if he could share his thoughts, he got very quiet, actually looking around to see if other hotel employees were close by. By then, we entered the elevator that had arrived. While subsequently walking down the hall to my room, George relaxed and began to discuss his current boss.

Based on his body language, I wasn't surprised that he didn't have a lot of positive comments, including not being very comfortable whenever the boss came by. Evidently, the new boss didn't spend a lot of time in front of the hotel and didn't bother to learn the names of employees. He had actually terminated one of the bellmen within two weeks of taking the management position with no explanation provided. The bellmen weren't even comfortable speaking with him about guest concerns since they were afraid they would be blamed for the problem or possibly get fired.

While I didn't have the opportunity to meet the new boss, I imagine that he felt like he was doing a good job. However, it appeared that he was inadvertently alienating the very personnel that were the key to his success. Now to clarify, the comments from George regarding his new boss were based on his perceptions.

A word of caution: many managers, especially those new to their role, have a tendency to overlook or discount the importance of an employee's beliefs/perceptions. Conversely, when management recognizes that the employee's beliefs are real to the employee and management responds accordingly in a positive manner, it is a significant step towards an enhanced working relationship. Without this perspective, long-term challenges can be encountered. Examples include reduced morale, lowered productivity, and higher turnover of personnel.

○ **Large research laboratory:** During my diverse career, I have also assisted numerous research laboratories with options to sustain or enhance their current safety culture. Due to the broad spectrum of research being conducted, each of the laboratories had a very diverse workforce. Personnel included graduate students, world-renowned researchers, including Nobel Laureates, and very talented craft personnel. Even though the research complexes were comprised of numerous buildings across many acres of land, the facilities were very clean with equipment and tools appropriately stored when not in use.

Based in part on the work I had done with the management teams for an extended period of time regarding safety culture, they were very engaged with their personnel. Managers routinely spent time in the facilities, sat in on staff meetings, and joined the workers for lunch in the laboratory cafeteria.

During my time at one of the research laboratories, I got to know numerous personnel, ranging from senior managers, researchers, safety specialists to security guards and maintenance personnel. A constant that was readily visible throughout the organization was their commitment to safety, working as a team, and helping ensure mission success.

Whenever I had the opportunity, I would seek out Charlie, a member of the janitorial team. Charlie had been at the laboratory for over 40 years and was routinely sought out by younger members of his group looking for advice on how to perform their work. He always had a kind word to share and a smile on his face, and he took great pride in his work.

On one of my visits to the laboratory, I noticed that Charlie had placed a piece of long clear plastic tubing into the suction side of the small vacuum cleaner he was using. After switching on the vacuum cleaner, he

1.4: How Do I Know if I Have a Good Safety Culture?

would then slide the other end of the clear plastic tubing into threaded inserts that were set flush with the concrete floor surface.

I had seen bolts placed into these threaded inserts so that research equipment could be secured to the floor to prevent movement or vibration while the equipment was in use. What really resonated with me was that Charlie also realized what the threaded inserts were being used for and had been inventive enough to come up with a way to clean the threads inside the inserts. When asked, Charlie explained to me that he knew that those inserts had to be free of dirt so that the "research fellas" (his words) could make sure their equipment didn't move around. Not too bad for an individual approaching 70 years of age who hadn't graduated high school.

> When asked, Charlie explained … so that the "research fellas" (his words) could make sure their equipment didn't move around. Not too bad for an individual approaching 70 years of age who hadn't graduated high school.

By now, hopefully you are gaining an enhanced understanding as to why an effective safety culture is critical to the success of any company. The next chapter addresses the vital role that you, as a leader, must play in ensuring that this commendable objective can be achieved.

Vision is the art of seeing things that are invisible.

– Jonathan Swift

2.0

LEADERS

Like most of us, I can recall having to work for a variety of managers with different leadership styles. Some of the managers made me feel part of the team, others, not so much or not at all. In most cases, when employees identify with a "good" manager, they are commonly connecting with the behavior (e.g., style) of the leader. One of the interesting things about leadership styles is that leaders who have the most flexibility in their style are more capable of getting the best from their employees.

Another aspect that can be readily overlooked is the power of humor and willingness to convey heartfelt emotions (Daniel Goleman, Richard Boyatzis, Annie McKee, 2013). This includes striking the balance between being composed/self-aware and having high energy, being motivating and emphatic. While a series of these topics can prove difficult for some leaders to emulate, I strongly recommend examining options to address them.

By way of example, during the summer of 2018, I was selected to conduct an independent evaluation of the safety culture for a large company located in the Northwest. The company, which has approximately 2,000 employees, had recently gone through a change in senior management, and their client had expressed a series of concerns regarding injuries and ineffective corrective actions. To gain additional insight, I spent two weeks at the site, which included meeting with craft personnel on the night shift and weekends.

While the company had a series of sound safety culture attributes in place, one area that proved challenging was senior management engagement. This was exacerbated by the personality of the new president when compared with the individual (Oscar) who had previously served in that role. Oscar was a very gregarious individual, and I imagine you have had the opportunity to meet someone with similar traits. Oscar never met a stranger, was the first person to walk up and shake your hand, made you feel your opinion was valued, and that you were considered part of the team. It should come as no surprise that Oscar's working relationship with craft personnel was particularly noteworthy. The craft welcomed him with open arms and looked forward to having him visit their work locations. As an aside, one of his many strengths was an insistence that any and all concerns or improvement suggestions provided by the craft be acted upon. This included ensuring that an explanation was provided to the individual who originally initiated the discussion.

Unfortunately, the new president (Andrew) had a completely different personality. Andrew was fairly introverted and was somewhat uncomfortable when meeting people at the jobsites. He was also very concerned about the relationship with the customer as well as current market share. In other words, he had a lot of "stuff" on his mental plate. Collectively, this contributed to a perceived persona of a lack of interest in engaging with his personnel. The results of my independent evaluation, including challenges regarding Andrew's engagement with workers, were captured in a detailed report that captured strengths, opportunities for improvement, and stressors.

When I returned to the site to present the results of my independent evaluation, Andrew requested that I meet with him separately after completing my presentation. During that subsequent meeting, he expressed concern regarding the section of my report that addressed the personality differences between him and Oscar. This discussion proved very insightful for Andrew, and we collectively identified a series of strategies that he agreed to implement to enhance his relationship with company personnel. For additional insight regarding some of these strategies, refer to Section 2.3, Management by Walking Around (Versus Just Stumbling About).

Similar to the initial discussion in this book regarding your safety culture not necessarily being good or bad, the same can be said for leadership styles. There are numerous publications addressing leadership styles, including those by Thomas Krause and Kristen Bell (Thomas Krause and Kristen Bell, 2015). To provide a starting point, the following three general examples are included in this book. These types of leadership styles, or similar versions thereof, can be encountered by the majority of employees in the workplace.

Authoritarian: This type of leader attempts to closely control the work of their employees. Motivation commonly consists of threats and discipline. An authoritarian leader is driven by fear of losing their "power," and the employees can be viewed as threats. Decisions are commonly rendered without any consideration of potential impact to the employees, and information supporting the decision-making process is very rarely shared. Under this type of leadership, employees have limited, if any, opportunities to learn and grow.

With this type of leadership, it is not uncommon for employees to feel frustrated, angry, and disengaged. Even more troubling, they can quickly lose their enthusiasm and have little interest in supporting new initiatives. Ironically, the leader can also become frustrated due to the intense day-to-day direction that is required, including feeling that everyone must be watched as well as having to approve all decisions. This extreme method of micromanaging can ultimately result in high employee turnover (e.g., attrition). At some point, it is not uncommon for the leader to ultimately be replaced.

Democratic: This type of leader is comfortable with shared leadership. Members of the group take a more participative role in the decision-making process. In the majority of instances, everyone is provided the opportunity to engage in open dialog with exchange of ideas being genuinely encouraged. While the democratic process commonly focuses on group equality and the free flow of ideas, the leader remains highly engaged so as to provide guidance and maintain overall decision-making authority.

When consistently applied, this type of leader inspires trust and respect. This is due to their sincerity and their decisions being based on their morals and values. In turn, members of the group/company commonly feel inspired and willing to remain engaged, including being an active contributor. This type of leader also tends to seek diverse opinions and does not try to silence dissenting voices or those that offer a less popular point of view. By definition, these leaders commonly realize the most success in engaging with their workforce.

Laissez-faire: This type of leader is the opposite of authoritarian. This type of leadership style has been described as management by withdrawal. They routinely delegate decisions to others, including what is considered acceptable standards for quality, work ethic, disciplinary actions, etc. This type of leader also has an almost uncontrollable desire to be liked. In turn, this feeds their hesitation to be demonstrative with their decisions since not everyone may like them. Unsurprisingly, this technique commonly results in employees becoming frustrated since it is not clear what is important to their leader.

Ironically, it is not uncommon for this type of leader to eventually turn on their employees. They can become cynical of the very individuals who are crucial to the leader's long-term success. In the current competitive environment, this type of leader commonly does not survive. By way of example, a company I was supporting had brought a new president on board. After a few months in his new role, he determined that a sweeping reorganization was required.

After approximately 18 months (that's correct, one and one-half years), the reorganization was finally published. His explanation for the delay was that he had to ensure everyone's needs were addressed. Due to the lack of clear roles and responsibilities for some of the newly created positions, additional stress was introduced into the company. The president was offered the opportunity to retire shortly thereafter.

In the majority of instances, the most appropriate leadership style is dependent on the situation and personality of the organization. By way of example, I have participated in numerous emergency management exercises/drills. In that situation, leaders need to have more of an authoritative approach versus continuing to solicit everyone's opinion.

One of the keys to being an effective leader is adapting leadership style as needed to address changing work environments, operational tempos, personality types, and similar considerations. Personality types that can be encountered in the workplace are addressed in Chapters 3 and 4.

An additional topic that has become more visible over the last few years is the importance of recognizing how the ego of the leader can contribute to the decision-making process. In the majority of instances, decisions based solely on the leader's ego will not prove beneficial long term. In other words, being able to recognize ego driven decisions versus those that are based on factual information and responding accordingly can greatly assist any leader during their safety culture journey.

Similar to the three leadership types discussed above, there are also many opinions regarding what makes a good leader. One broad based study offered a series of traits that should be considered. Examples include possessing the ability to identify which practices need to be eliminated, having the right people on the bus, and an earnest commitment to placing success of the organization above themselves (Jim Collins, 2001). Regardless of your actual leadership style, a few topics consistently float to the top.

— **Vision**: You have to know where the company is heading and how you're going to ensure the company can get there.

— **Commitment**: You need to remain committed to your vision. While you obviously want to listen to feedback from your senior team, at the end of the day, it is ultimately your decision.

— **Situational Awareness**: You need to anticipate changes that can impact your company and respond accordingly. This may include needing to adjust corporate strategies in anticipation of changing financial and/or business sector climates.

— **Strength**: Even when you may lack confidence (it happens to all of us at some point), you must be able to still portray the strength that others can look up to and respect.

— **Grit**: This is personally one of my favorites. There is a wide variety of synonyms in dictionaries that address grit. Examples include courage, steel, mettle, pluck, resolve, tenacity,

> ... a leader with grit has the ability to continue on, regardless of the challenges encountered.

perseverance, and endurance. To summarize, a leader with grit has the ability to continue on, regardless of the challenges encountered.

From my perspective, a great example of grit was the response by New York City Mayor Rudy Giuliani after the terrorist attack on September 11, 2001. Like most leaders, some of his decisions were not fully embraced by everyone at the time. Regardless, he was able to provide a consistent presence and ensure decisions were executed. Mayor Giuliani was responsible for coordinating the efforts of a broad spectrum of city departments and interfacing with state and federal entities, while juggling a series of discussions with media and public appearances.

Perhaps one of his greatest strengths was his ability to connect with New Yorkers and mirror their heartfelt emotions of shock, anger, and commitment to rebuild. This ability to connect was reflected in a poll after the attack, where his approval rating jumped from 36% to 79% (Anthony J. Viera, Rob Kramer, 2016). During a subsequent 9/11 memorial service, Oprah Winfrey affectionately referred to him as "America's Mayor."

Compare the capability of Mayor Giuliani to be embraced by his constituents with the response received by Mr. Tony Hayward when photographed while sailing his 52-foot yacht during the Isle of Wright race. While competing in this type of yacht race was not unusual, Mr. Hayward chose to be on the yacht within days of appearing at a Congressional committee hearing addressing the Deepwater Horizon catastrophe. This is the same accident addressed in Chapter 1 of this book.

To compound this massive public relations nightmare, Mr. Hayward was quoted by numerous news sources a few weeks before the yacht race, "There's no one who wants this over more than I do. I would like my life back." Numerous individuals, including family members who had lost a loved one during the Deepwater Horizon explosion and fire, were very agitated with Mr. Hayward's comments, stating they would like to get their lives back as well. Rahm Emanuel, Chief of Staff to President Obama, probably summarized this situation best by concluding that Tony Hayward was not going to have a second career in public relations consulting.

Now this is where it gets rather interesting. Being able to mesh leadership attributes with having the ability to truly connect with personnel at the task level is essential but can prove problematic. As noted earlier in this chapter, when viewed from a safety culture perspective, leaders who have the capability and courage to adjust their leadership style for the betterment of their company commonly realize the optimum success.

This topic is rather fascinating since, by definition, the same individual who has to demonstrate leadership also needs to be able to mentally step back and let others take the lead when attempting to learn what is really occurring in the workplace. This will be discussed further in subsequent chapters.

A leader is one who knows the way, goes the way, and shows the way.

– JOHN C. MAXWELL

2.1

Isn't Just Being the Boss Good Enough?

Not that many years ago, employee injuries were routinely blamed on the individual worker, often referring to the worker as "accident prone." Unfortunately, the concept of employees being accident prone still resides in some organizations to this day. I have even encountered companies referring to website articles claiming that over 80% of injuries are the fault of the worker. Needless to say, I become extremely vocal when I encounter this type of sentiment within a company since this "urban myth" is an insult to any workforce and does not reflect reality.

> Not that many years ago, employee injuries were routinely blamed on the individual worker, often referring to the worker as "accident prone."

Not surprisingly, while attempting to recover from the Great Depression, the United States faced massive unemployment, and workers were desperate to have any job so that they could feed their families. In other words, workers were viewed as being expendable. Collectively, numerous companies had taken a "do it because I say so" strategy.

2.1: Isn't Just Being the Boss Good Enough?

This attitude was readily evident during construction of the Hoover Dam, which was dedicated in the fall of 1935. During the construction phase, the construction company owners denied requests from workers to have fresh water provided each day along with periodic breaks in the shade during each shift. Bear in mind, the workers were constructing the Hoover Dam outside of Las Vegas, Nevada, with summer temperatures approaching 110 degrees. Since there were hundreds, perhaps even thousands, of people waiting outside the construction gate looking for work, the owners decided that there was no need to accommodate requests regarding work conditions.

Gradually, the tide began to shift when companies began to realize the critical role that management plays in day-to-day operations. The need for management to own their processes was also recognized by the Occupational Safety and Health Administration (OSHA). This regulatory agency has issued numerous fines and penalties for workplace hazards that had not been effectively abated. Those fines and penalties are commonly adjudicated at the company level versus at the individual level.

Obviously, no one can be everywhere in their facility at any one time. When using a heavy-handed management approach, it can prove difficult to ensure personnel do what is expected of them when the boss is not present. This is somewhat akin to how well drivers comply with the posted speed limit on the nation's highways.

In many instances, it is not uncommon for drivers to exceed the posted speed limit by 10–20 miles per hour or more. They justify this practice by being confident in their driving abilities, wanting to save time, or being late for an appointment. However, if a highway patrol or police car is observed in the immediate area or parked on the side of the road operating a radar detector or issuing a citation, these same drivers will immediately reduce their speed to avoid a ticket. In other words, many drivers will only follow the posted speed limit if law enforcement is present.

Replace the speed limit scenario described above with the employees at your company. When companies rely on a "do it because I say so" approach, they are setting themselves up for failure. In essence, your company has to become the patrolman/policeman attempting to enforce safety practices. To address this challenge, some companies direct their safety professionals to serve as "safety cops," issuing citations when workers do not follow safety rules. Similar to the speeding scenario, this approach commonly does not provide any long-term value and can inadvertently generate friction between your employees and your safety professionals.

During my career, I have also provided senior safety consultative support at commercial nuclear power plants across the United States. My support included assisting a plant in southern Florida prepare for final licensing inspections by representatives of the U.S. NRC. While coordinating a walkdown of the reactor facility with senior management, we encountered a great (or perhaps horrific) example of the "do it because I say so" mentality.

We were shocked to realize that a series of "azimuth" markings on the inside walls of the building were no longer visible. These special markings are scribed into the concrete surface at approximately eye level and are utilized to confirm equipment location and configurations. These markings, highlighted with fluorescent paint for increased visability, are also verified by NRC and plant representatives during routine walkdowns and subsequent inspections.

Upon further investigation, it was determined that a new foreman for the paint crew had told his crew members to clean and paint all surfaces. The work was already behind schedule, and the foreman wanted to make

2.1: Isn't Just Being the Boss Good Enough?

a good impression with his boss. When asked by the crew about the azimuth markings, the foreman scolded them for asking too many questions and told them that if they didn't want to do the work, others would. In other words, do it because I said so.

While the foreman was quickly removed from the job, additional costs were realized due to the rework. Special survey crews had to be brought in over a weekend to confirm location of the now visible azimuth markings. Confirmation checks then had to be conducted by quality control personnel. In other words, a very costly and time-consuming process.

Prior to attending college, I had a series of jobs in the Midwest, including working at a large plant that bottled soft drinks. When I eagerly showed up for work on the first day, I was told to report to the warehouse, where there were approximately 25–30 people working. I met Fred, who was the boss for the warehouse. He was fairly curt with me while providing initial work instructions, which included telling me to work with some of the older guys until I knew what was going on.

While this wasn't necessarily unusual, what I remember most was him not being interested in learning my name. In fact, he was very candid, telling me that most guys "my age" don't last more than 2–3 weeks, and he wasn't going to waste his time trying to remember my name. So … he simply referred to all the new guys as Joe. When you were a new guy working in the warehouse and you heard Fred scream "Hey Joe," you were expected to stop what you were doing and look at him. He would then point to which "Joe" he wanted.

Needless to say, there was very poor morale, and the plant had fairly high personnel turnover, especially in the warehouse! While this may be viewed as an extreme example given today's work environment, some companies still don't have a true appreciation of the importance of having an engaged workforce. This topic is further discussed in later chapters of this book.

Before leaving this topic, I want to point out that this situation can be encountered in large professional organizations as well as small companies. A few years ago, I was asked to spend approximately six weeks with a team of researchers who had been tasked with developing presentations that would be shared with an oversight group. The oversight group, who reports to the United States Congress, was tasked with conducting independent reviews of research activities. In addition, this oversight group also served in a key role regarding the ability for this team to continue their work as well as receive additional funding. In other words, these presentations were critical to the lifeblood of the research team.

The presentations had to address how the researchers ensured that safety was built into their processes, including the need to provide a healthy safety culture for the personnel supporting their efforts. Based on prior support I had provided to the laboratory manager, I was asked to step away from other contract commitments and focus solely on this effort. This included leading a series of brainstorming sessions that dealt with the expectations of the laboratory manager in addition to subsequent meetings with the researchers addressing the overall process.

Among the topics discussed was the "elephant in the corner." In other words, if I was aware of an issue that was not addressed in the draft presentations, it was incumbent upon me to point it out. Early on in the process, I also had a series of additional group sessions with the presenters. Topics included overall expectations, as well as the need to ensure potentially sensitive topics (e.g., current safety culture challenges) were appropriately addressed in their respective presentations.

2.1: Isn't Just Being the Boss Good Enough?

Over the next few weeks, I met one on one with each of the researchers who would be serving as a presenter. I also spent time in their research laboratories, which provided me with additional insight regarding daily activities. This also helped me to appreciate the "atmosphere" that might be encountered by the oversight group during their pending visit.

As the preparation efforts neared completion, each of the researchers conducted a "dry run" (e.g., rehearsal) utilizing the slides they had developed. To provide maximum value for the effort, each presenter walked through all of his/her slides, including supporting discussions, exactly the same as those that would be shared with the oversight group. I intentionally requested this approach be utilized so that the laboratory manager and I could gain a sense of how the overall message was going to be conveyed as well as ensure the overall presentation was on target. The series of dry runs were completed over a three-week period, and the majority of the draft presentations met expectations. However, one presentation proved problematic.

That presentation was conducted by one of the lead researchers (Theodore), with his support team in attendance. By the way, Theodore was a senior level PhD who had a reputation of being very opinionated. He was also not very receptive of thoughts or opinions that did not agree with his. Some support personnel had actually requested to be reassigned to a different researcher because they were not very comfortable having to work with him.

Theodore had over 40 slides that addressed different types of research he and his team had conducted, including numerous pictures of a very complicated series of components. At the end of his presentation, he stated that he was very confident that the presentation was ready to be published and concluded his thoughts by noting that unless we had any questions, he was going to leave for the day.

The laboratory manager thanked Theodore and his team for their efforts, and his team appeared to be very relieved that the presentation was completed. After stepping out into the hall, the laboratory manager expressed some of his concerns to me regarding the presentation he had just observed. The laboratory manager then asked that I lead the discussion with Theodore when we returned to the conference room.

While I had been warned in advance about his somewhat arrogant attitude, I also appreciated the importance of having a presentation that addressed the agreed upon criteria. With that in mind, I initiated my critique by gently reminding Theodore of the overall objective as well as my role of providing an independent perspective, including having to be completely candid. I then proceeded to provide feedback that addressed some of the technical topics addressed in his slide set.

Knowing full well that I needed to get his attention, I summarized my remarks by joking that I definitely was not a researcher and that some of his devices looked like very shiny, very expensive toaster ovens. I then concluded that while his work contained in the slides was very impressive, when it came to addressing the safety of his team and his personal commitment to same, it appeared that he couldn't find his behind with a mirror and a flashlight! To help smooth the waters so to speak, I then went on to state that while I was confident about his personal commitment to safety for his team, it simply wasn't being reflected in the draft slides.

Since Theodore wasn't used to being addressed in such a manner or having his perspective challenged, he became very frustrated. To coin an old phrase, you could have heard a pin drop in the room. Every member of Theodore's team actually held their breath awaiting his response to my comments. The team members even went so far as to avoid eye contact with Theodore! After a few moments, he glared at his team and asked if

2.1: Isn't Just Being the Boss Good Enough?

my perspective was correct. After some hesitation, the team members cautiously nodded their heads. The laboratory manager also concurred with my evaluation.

To his credit, Theodore paused and then became very passionate while stating that nothing was more important to him than the safety of his team. After he gained his composure, he grudgingly acknowledged that his presentation had been developed using his traditional strategy for sharing his research processes and associated results. The breakthrough came when he asked me to assist him with the revision of his presentation so that it would clearly address his commitment to safety. During a series of subsequent one-on-one sessions, both of us gained a better sense of each other's skills, including our mutual commitment to safety as well as mission success.

Perhaps the greatest insight was gained when Theodore paused during one of our sessions, asking if I had any insight as to why his team had not initially pointed out what was missing in his presentation. Before answering, I asked him if he had any thoughts regarding what may have contributed to that situation and if it had occurred before. That question turned out to be a watershed moment for him as he began to talk about his internal frustration with some young researchers not readily sharing their opinions or not wanting to work with him. After a series of discussions, the amount of personal insight he gained was incredible, ultimately culminating in a significant change regarding how he and his team interacted with each other.

Ultimately, he was very complimentary to me about having the courage to voice my opinion and assist him with taking a fairly harsh look in the mirror. While I really appreciated his comments regarding my ability, I firmly believe that he demonstrated the most courage by being willing to acknowledge and act upon the need to change his approach. By the way, over the next couple of years, young researchers began clamoring for the opportunity to become part of Theodore's team.

The "do it because I say so" approach simply becomes a waiting game for a failure to occur and how the failure will manifest itself. A given company may begin to experience an increase in employee absenteeism, excessive turnover, reduced quality of work products, and/or poor customer service. All of these undesired attributes can be viewed as an indicator of an ineffective safety culture that can ultimately lead to worker injuries and/or fatalities.

> The "do it because I say so" approach simply becomes a waiting game for a failure to occur and how the failure will manifest itself.

Another area that is overlooked is the negative impact due to worker fatalities/injuries/illness. By the way, after investigating a wide variety of workplace injuries, I have never encountered a scenario where the employee intended to hurt themselves. With few exceptions, your employees take great pride in performing their work and enjoy the contributions they make to the company's success. As noted at the beginning of this chapter, unfortunately there are still some companies that only focus on worker error or are more than willing to simply close out the investigation due to the worker being accident prone. These types of approaches reinforce a horrific safety culture that stifles open communication and presents significant barriers to a partnering relationship between management and the employees.

In addition to the emotional impact to the worker, their family, and their peers, these incidents can also result in increased insurance rates (based on Experience Modification Rate [EMR] calculations), reduced productivity, and schedule delays. In some instances, a company can be prohibited from bidding on new projects or contracts due to having an

2.1: Isn't Just Being the Boss Good Enough?

EMR score that is higher than permitted. In turn, this can reduce company profits and potential for company growth.

Refer to Chapter 5.2, Metrics, for additional insight regarding OSHA injury rates. This chapter also addresses challenges that can be encountered when attempting to rely solely on those rates to determine overall safety and health performance.

Finally, to paraphrase an old saying: "You don't know what you don't know." I have had the pleasure of providing safety support to a wide variety of private sector companies and federal employees and their contractors for over thirty years. To date, I have yet to meet any company president or member of senior management who was willing to knowingly accept these risks. Candidly, since you have chosen to read this book, you already appreciate the fallacy of attempting to operate your facility/company with the "because I said so" approach as well as recognizing the importance of having an engaged workforce.

The growth and development of people is the highest calling of leadership.

– Harvey S. Firestone

2.2

THE IMPORTANCE OF INVOLVING PERSONNEL

The importance of employee involvement cannot be over emphasized. Very few, if any, companies have been able to sustain long-term implementation of initiatives without the support of their employees. Engaging employees also provides valued insight regarding how work is actually being performed. In other words, what is really going on day-to-day.

> Very few, if any, companies have been able to sustain long-term implementation of initiatives without the support of their employees.

Conversely, not involving personnel will inevitably lead to subtle "push back" when the initiative is being rolled out. Ensuring task level personnel are selected to serve on the team(s) gains a series of benefits, including increased buy-in. Subsequent chapters of this book provide a series of perspectives regarding composition of teams when developing initiatives.

2.2: The Importance of Involving Personnel

By way of example, a few years ago I was tasked with leading a team responsible for the development of a large document addressing how a company was implementing new safety criteria specified by federal law. If the deliverable was not completed in a timely manner while also ensuring a high level of accuracy, the company could be subject to significant fines and/or potential contract suspension.

This effort had to address numerous topics, including electrical safety, worker safety, fire protection, construction safety, and occupational medicine. Current company procedures also had to be reviewed to determine what revisions were required. A detailed project schedule had to be developed with the overall process taking approximately three months to complete.

Due to the breadth of topics that had to be addressed, as well as the examination of the implementation of associated procedures, I ensured task level personnel were also selected to participate in the working group. The group was comprised of approximately 25 personnel, including representatives from safety and health, fire and rescue, human resources, training, and project management. Realizing the unique perspective that task level personnel offered regarding actual implementation of criteria, I also ensured they were afforded the opportunity to actually write portions of the document.

Some members of the senior management team weren't initially comfortable with this approach since they felt that the workers really didn't have a lot to offer and didn't have the ability to serve as authors for portions of the document. Ironically, some of the task level personnel initially felt the same way regarding not having the skills to complete the task.

These concerns were addressed through use of a series of sub-teams, with at least one task level individual on each team. I also facilitated a kick-off meeting, followed by periodic (e.g., weekly) meetings that addressed progress. During the kickoff meeting, one technique utilized was to ask each member to discuss options regarding how to approach the task, along with how they could assist. Within a fairly short period of time, the task level personnel were very comfortable writing portions of the document as well as performing "peer reviews" of text generated by other sub-teams.

The final deliverable not only readily met the objectives, it also served as a template that other companies utilized to address the same criteria. More importantly, the task level personnel on the working group helped spread the word throughout their respective organizations regarding the importance of their efforts, which greatly enhanced ultimate buy-in and long-term implementation. One unexpected benefit was the task level personnel gained an appreciation for the effort required to actually prepare such a large deliverable. Prior to this assignment, they thought the "managers" simply went to meetings or drank coffee all day.

Based on this success, a similar process was utilized when this same company chose to pursue Voluntary Protection Program (VPP) Star certification. This certification is awarded to companies that can demonstrate high levels of management commitment, employee engagement, as well as reduced injuries.

An external VPP Audit Team spent two weeks at the company, reviewing procedures, observing work practices, and interviewing personnel. The company readily received the VPP Star certification, and the external audit team was very complimentarily regarding the high level of participation by task level personnel.

Candidly, the topic of employee involvement often gets overlooked when companies are developing programs, writing procedures, and conducting audits and accident/incident investigations. The importance of this topic has been addressed by a series of notable authors, including Aubrey Daniels (Daniels, 2016). As provided in Chapter 3, there is a wide variety of approaches, each with its own series of challenges and benefits.

Research indicates that workers have three prime needs: interesting work, recognition for doing a good job, and being let in on things that are going on in the company.

– Zig Ziglar

2.3

Management by Walking Around
(Versus Just Stumbling About)

As previously discussed, the task level personnel at your company truly understand how work is being performed, what improvements could be made, and what problems need to be addressed. The majority of these individuals has been working at the facility for an extended period of time and has seen a wide variety of "feel good" initiatives come and go. It is not uncommon to hear comments along the lines of "flavor of the month," "another 30-day wonder," or similar thoughts.

Another attitude that can be encountered is NIMBY, which stands for Not in My Backyard. In other words, the employees will quietly not support the new initiative, knowing that in the majority of instances, the initiative will fade away in the near-term. Unfortunately, in many cases, the company confirms this perception by not providing a long-term commitment for the new process, etc. In turn, it becomes that much more difficult to roll out subsequent initiatives, which can generate additional friction between the senior team and the workforce.

2.3: Management by Walking Around

When I hear these types of phrases, and I've heard them a lot during my career, or encounter these situations, I viewed them as being reflective of an organization that has not truly engaged their workforce prior to rolling out a new process. This is not that unusual to encounter since many companies attempt to roll out an initiative without appropriate planning, hoping for a quick win. Refer to Ready, Aim, Fire discussion contained in Chapter 4.1 for additional insight regarding this challenge.

One of the best methods to overcoming this challenge is to be "on the shop floor." In other words, spending time with personnel in their actual work environment. Although this requires you to be away from your office and some meetings may have to be rescheduled, the results of these "walkabouts" can prove to be well worth the effort.

Now this is where it gets really interesting. As noted earlier in this chapter, the same attributes that can be an asset as a president can prove to be a hindrance when attempting to engage with workers in the workplace. Hopefully, the following discussion provides additional insight.

> … the same attributes that can be an asset as a president can prove to be a hindrance when attempting to engage with workers in the workplace.

In working with numerous companies and their management teams, I have identified a series of consistent themes/challenges that have been encountered when attempting to roll out a Management by Walking Around (MWA) campaign. In some instances, the end result can be better described as management by stumbling about. This shouldn't be viewed as belittling the efforts of senior management. It is simply a recognition that when senior management attempts to engage with the workforce in their work environment without a comprehensive strategy, there is the potential to generate more harm than if not attempted at all.

One company I worked with (given the fictitious name AXXX Services) decided that senior management needed to be "in the field" more often. The overall concept was very well intentioned. Namely, make a focused effort to spend time interacting with the workers in their work environment versus just during routine meetings. By the way, this approach has proven very beneficial for numerous companies. Examples include heightened communications, improved relationships, increased comradery, and enhanced safety culture. Anecdotal discussions with company representatives indicated that they also realized increases in productivity, reductions in absenteeism, and lower injury/incident rates.

Based upon what they heard about other companies' successes, AXXX Services launched their own MWA Initiative. This initiative included requiring performance of at least two "field visits" each month, with the results documented using a three-page form. The form included an exhaustive checklist that contained a series of "boxes" that needed to be marked regarding problems identified, facility issues that needed to be corrected, etc.

Naturally, a new procedure had to be written. Since the company wanted to get the procedure issued quickly, there was a very limited review and approval cycle. In other words, many of the personnel who were going to be tasked with implementing the procedure were not provided the opportunity to participate in its development. As a result, the procedure wasn't very well crafted, which contributed to a myriad of implementation challenges.

To further demonstrate how seriously this initiative was being taken, the president required that the number of field visits be reviewed during his weekly meeting with his management team. However, there were no actions taken against senior managers who did not meet the goal of two field visits per month (e.g., no accountability). Even more troubling,

2.3: Management by Walking Around

while the president emphasized completion of the field visits, there was no discussion during the meeting regarding what was actually learned from the time spent in the field.

Perhaps the greatest irony was that the completed field visit forms were not reviewed by anyone after they were submitted via email. I repeat, nothing, absolutely nothing, was done at the company level. This generated additional confusion and frustration by the managers who were actually conducting the mandatory field visits. Not surprisingly, the majority of these individuals reverted to simply filling out the form without any focus being placed on enhancing their working relationships with personnel in the workplace. This initiative became more convoluted since some senior managers didn't think it applied to them because they only had people who worked in an office environment.

Similar to a lot of company processes that are not well thought out, their MWA Initiative slowly faded away as senior management became distracted with the latest issues, new initiatives, etc. Ironically, while the management team of AXXX Services emphasized reviewing injuries, missed deliverables, budget overruns, etc., there was no emphasis placed on why their MWA Initiative was not successful.

Conversely, I had helped companies roll out MWA Campaigns that ultimately enhanced the overall organization. While there are many approaches, here's a strategy that I've found works well.

— **Clearly Defined Objectives**: While this should be self-evident, personnel throughout the organization need to understand the purpose of the campaign. The objectives should also address how individuals can assist with the effort as well as what will be done with the results.

— **Ability to "let go"**: Managers need to appreciate that they may not have a lot of insight regarding some of the tasks they are observing.

In other words, these managers need to be willing to listen much more than they talk. This supports a successful MWA since decisions commonly do not need to be immediately rendered. This is a foundational concept that needs to not only be acknowledged but embraced. Having said that, I fully appreciate how difficult this can be for some managers. However, by making a concerted effort to let others lead the discussion, the results can prove immensely valuable.

— **Campaign Facilitation**: Having someone facilitate initial roll out of the MWA campaign can greatly assist with addressing potential challenges. Ideally, this individual has some level of experience interacting with task level and craft personnel. This relationship can also provide consummate value with overcoming hesitancy that may be initially encountered.

— **Organic Approach**: This is one topic that can inadvertently get overlooked. I use the term organic in recognition of the need for the managers to appreciate that they need to be flexible and adapt to what they encounter during the field visits. This philosophy will also assist with continuing the MWA campaign even though results of initial efforts are less than ideal.

This shouldn't be a surprise since task level personnel may not have encountered the managers in their work locations previously and are commonly somewhat hesitant to readily engage in open conversations during initial MWA visits. In turn, the manager conducting the MWA may feel that it was a waste of time because they could have remained in their offices, attended routine meetings, etc. Having the ability to "stay the course" is essential to long-term success. In this case, success translates to building camaraderie with personnel closest to the work.

— **Use of a "Tour Guide"**: In the majority of companies, there are lower tier levels of management (e.g., foremen, superintendents) that can

2.3: Management by Walking Around

help facilitate the MWA at the field level by accompanying the managers when the MWA is being conducted. Another group that can readily serve in this capacity is safety and health professionals. This is somewhat akin to having a tour guide when you are on vacation.

When my wife and I would travel to Jamaica (beautiful island by the way), we routinely requested a tour guide. They had great insight regarding local activities, maximized our time on the island, and helped us appreciate the local culture. We also found that since the locals knew our tour guide, they were more willing to chat with us. In other words, we weren't just another pair of tourists. In a similar manner, selecting someone who has a strong working relationship with task level and/or craft personnel can significantly enhance initial MWA efforts.

By way of example, I recently assisted senior management of a company with their MWA efforts. While touring a large warehouse with their president, I was approached by one of the craft personnel (Sammy) regarding what was going on and why the president was in "his" area. I had known Sammy for many years and spent a lot of time talking with him about his Harley Davidson motorcycle as well as different recommendations he had regarding how to improve warehouse operations.

At approximately 6'5" in height and weighing in excess of 275 pounds, Sammy was an imposing figure. However, he was clearly rattled with the president being in the warehouse. I took a few minutes to explain what was going on, including the president looking forward to speaking with Sammy. At that moment, Sammy looked me squarely in the eye and asked if the president was "cool." Upon being assured by me regarding the "coolness" of the president, both of them ultimately had a great conversation together. Sammy eventually got so comfortable that he invited the president to join him on a motorcycle ride, but I'm not sure if the president ever took him up on that offer.

Through implementation of this tour guide approach, I've seen companies experience significant improvements regarding their overall working relationships with task level personnel. As noted previously, one group that can provide great value is your safety and health personnel who are stationed (e.g., deployed) in the field. In the majority of instances, these individuals have great working relationships with the craft personnel. Similar to the prior discussion regarding Sammy and the president, your deployed personnel can readily assist with introductions. This approach can also provide the opportunity for these deployed individuals to have some "face time" with senior management that commonly does not occur.

Insight gained from spending time in the field can assist other processes, including routine meetings. Examples include use of a President Safety Council (or equivalent) addressed in Chapter 3 to enhance communications with task level personnel. I have participated in a series of these meetings and can personally attest to the added value.

At one of these meetings that had numerous craft personnel in attendance, I was asked to serve as facilitator. Throughout the meeting, craft personnel talked about recent work experiences, activities that were going well, and potential process improvements. During a series of these discussions, I intentionally asked some of the craft personnel about activities going on in their lives outside of the workplace. Examples included restoration of old cars (e.g., hot rods), having their youngest daughter compete in karate competitions, and preparing for barbeque competitions.

While these individuals had been actively engaged in the meeting up to that point, the change in their persona when asked to talk about "non-work" activities or interests was particularly noteworthy. Each one of them would "light up" when talking about their children, hots rods, etc. In turn, this provided another avenue for finding more common

2.3: Management by Walking Around

ground among participants. In numerous instances, members of senior management had similar interests, whether it was working on cars or enjoying great barbeque.

This technique worked very well, and a series of senior level managers pulled me aside after the meeting. They were quizzical regarding how I knew so much about the craft's non-work activities. This question initially definitely took me aback a little. I then graciously explained that I had intentionally set aside time to chat with the craft at their work locations, including getting to know them as individuals. Again, while these strategies are not rocket science, I have found senior management needs to be periodically reminded of the value gained from applying them.

A great illustration regarding the concept of leaders being able to relate to their teams on an individual basis was provided by Brandon Black. Before retiring in 2013, Mr. Black served for nine years as President and Chief Executive Officer for Encore Capital Group. Among the core services of this publicly traded financial services company was acquisition of unpaid consumer debt.

During his time at Encore, Mr. Black was faced with a series of tumultuous events, including loss of financial share, scrutiny by consumer groups, as well as having to orchestrate a reduction in force (e.g., layoff) which eliminated approximately 10% of a large workforce. After the company eventually recovered and became sustainable, Mr. Black determined it was time for him to retire so that he could spend more time with his family and work with charitable organizations in his community. To assist with effective transition for the new president, Mr. Black agreed to remain with the company for an additional 90 days.

During that transition period, Mr. Black made the intentional decision to visit numerous personnel in their offices, sit in routine staff meetings, etc. Throughout this time period, Mr. Black received a series of accolades regarding how much personnel enjoyed working with him. At numerous work locations, the personnel had compiled a book containing thank you notes that was presented to him during his final visit.

Ironically, no one discussed his ability to be a visionary leader, to make key decisions quickly, or being smart and capable. Rather, the discussions revolved around his ability to get to know company personnel as individuals. Examples included remembering the names of their children, making people feel respected, and putting making the right decision ahead of being right. As noted in *Ego Free Leadership: Ending the Unconscious Habits that Hijack Your Business* that Mr. Black co-authored with Mr. Shayne Hughes, one of the thank you notes that Mr. Black considered most inspirational addressed his ability to connect with people and making sure everyone felt included, respected, and attended to (Brandon Black & Shayne Hughes, 2017). I am highly confident that any leader, including those reading this book, would be honored to be viewed in such a manner by their employees.

Based on the insight contained in this chapter, hopefully you are gaining an appreciation of the importance for you and your senior management team to have a strong working relationship with your employees. The next chapter addresses a series of considerations regarding employee involvement, including respect in the workplace, pride, and ownership.

Coming together is a beginning, keeping together is progress, working together is success.

– Henry Ford

3.0

Employees

The previous chapter addressed the importance of leaders and their leadership capabilities. As noted in the introduction of this book, employees also serve in a key role regarding safety culture processes. Throughout this book, there are numerous perspectives regarding employee engagement, including the importance of leaders being able to effectively connect with them.

A Gallup Poll conducted in 2015 identified that only approximately 30% of employees surveyed considered themselves "engaged" (Gallup, Incorporated, 2016). The majority of respondents (approximately 50%) responded that they were not engaged. Even more disturbing, approximately 17% of employees surveyed responded they were actively disengaged. These values have remained fairly constant, with little, if any, increase in overall scores from prior annual surveys. The 2017 Gallup *State of the American Workplace* Report identified little positive change (e.g., 33% versus 30%) regarding employees considering themselves engaged (Gallup, Incorporated, 2017).

At the risk of stating the obvious, members of the workforce who are actively engaged are the backbone of any successful company. These individuals are enthusiastic about and committed to their work. There is a strong correlation between engaged workers and key business outcomes, including productivity, profitability, and customer engagement. When you have employees who are engaged, they are instinctually in tune with potential innovation, can readily embrace change, and actively pursue process improvements.

While members of the workforce who are engaged are fairly easy to identify, the attributes of a disengaged worker can be more subtle. In many cases, these employees are not openly hostile or intentionally disruptive. While they may show up to work on time, they commonly will only put forth the minimum amount of effort to get by. These individuals are more inclined to call in sick, are not interested in working on special projects, and are very quick to "jump ship" when a new job opportunity appears on the horizon.

They can't wait for the next break, lunchtime, or end of shift so that they can go home. In some cases, they are more likely to steal from their company, have a negative influence on their peers, or reduce customer satisfaction to the point the customers seek other suppliers. Chapter 4.3 provides additional insight regarding these types of individuals.

Setting aside employee engagement/involvement for the moment, let's view some statistics from other topical areas of hypothetical company performance. Examples include:

— Only 30% of your employees arrive to work on time. This can translate to deliveries not being completed as scheduled, customers not receiving timely response to their questions, etc.

— Your company products have a rejection rate of over 70%. With many companies operating at fairly thin profit margins, the cost of refunds and/or rework can prove devastating.

— Retention rate at your company is below 40%. Stated another way, the cost to address attrition continues to increase. A good friend of mine owns a hot rod shop in Las Vegas, Nevada, and he estimates it costs approximately $15,000 to train a new employee. New hire costs for a large security company I've assisted were in the range of $100,000 to $125,000 per employee. Just to clarify, these costs are not commonly passed onto the consumer. In other words, these costs directly impact the bottom line/profitability.

— Employee complaints received by your Human Resource (HR) Department have increased by over 50%. It is expected that any company will have some employee complaints that need to be addressed by HR. However, if there is a significant increase, that could prove detrimental to the long-term health of the company. As an aside, there are commonly many more employees who are disgruntled beyond those who actually file a complaint.

— The value of your company stock has dropped over 65%. To quote Ricky Ricardo from the legendary long-running television series that he cohosted with Lucille Ball: *"Lucy ... you've got some 'splainin' to do!"*

I am highly confident that if you experienced any of the above hypothetical scenarios, much less a combination of them, you would immediately institute a call to action to determine the appropriate path forward. While that would be the logical response, the majority of companies do not place that same level of emphasis/concern regarding employee engagement. This is very puzzling since, as noted above, having employees who are actively engaged can directly contribute to the above scenarios not being realized. Conversely, employees who are not engaged can have a series of negative impacts as discussed earlier in this chapter. From a financial perspective, Gallup estimates that actively disengaged employees

3.0: Employees

cost U.S. companies approximately $480–$600 billion dollars per year (Gallup Incorporated, 2016).

When visualizing a company that can't successfully address employee engagement, it reminds me of a story regarding a new president selected to take over a company. He had the opportunity to meet with the retiring president, who congratulated him on being selected. During this meeting, the new president was presented with three sealed envelopes, numbered 1, 2, and 3 that he was to keep in his desk.

The new president was cautioned to only open the envelopes if the company began to experience problems that he wasn't sure to how to deal with. As with most transitions, things went smoothly for approximately the first year. Then, sales began to slow down and the Board of Managers (BoM) became concerned. Recalling the advice from the prior president, he went to his desk and opened the first envelope. There was only one sentence: "Blame your predecessor." The new president met with the board and, somewhat graciously, threw the prior president under the bus, so to speak. The BoM bought off on the story, and sales began to increase once again.

Unfortunately, over the next six months, the company started experiencing very high rejection rates for their products, and their social media account was flooded with complaints about poor customer service. Recalling how well the guidance contained in the first envelope had helped him, he immediately opened the second envelope while sitting at his desk. Instead of a brief sentence, that envelope only contained one word: "Reorganize."

Following that instruction, the new president directed a sweeping organizational change. He replaced mid-managers and fired some individuals he felt were contributing to the quality defects of their products.

Committed to demonstrating his ability to be a great president, he executed the majority of these decisions with little, if any, input from his senior team. Sure enough, overall sales began to improve once again, including a reduction in customer complaints.

While this would commonly be viewed as good news, the majority of employees were fearful that they could be the next ones to be fired, and therefore simply hunkered down until they could find employment elsewhere. Some of the mid-managers who were not replaced reverted to a heavy-handed approach when dealing with their direct reports, ignoring safety concerns, and attempting to discredit any complaints received by the HR Department. Due the lack of a response, numerous customers simply stopped submitting complaints and purchased their products elsewhere.

Unsurprisingly, performance of the company degraded even further, with no end in sight. In a complete panic, the new president rushed into his office and ripped open the third envelope. He was convinced the information it contained would ultimately save the company as well as his career. Imagine his dismay when he read these three simple words: "Prepare three envelopes."

Hopefully, you are never placed in the above situation. By the way, I am highly confident that the discussion and insights contained in this book can help you avoid the need to look for or prepare the dreaded three envelopes.

The topic of different personality types is addressed in Chapter 4, including the importance of being able to identify champions. That chapter also addresses personality types as defined by Tim O'Leary in his book entitled *Warriors, Workers, Whiners & Weasels, Understanding and Using the Four Personality Types to Your Advantage.*

While it is important to recognize different personality types, a given group and/or organization can have its own "personality" above and beyond that of an individual. The following discussion addresses the different types of groups/organizations that can be encountered in the workplace, including common responses to initiatives.

> While it is important to recognize different personality types, a given group and/or organization can have its own "personality" above and beyond that of an individual.

○ **Office Personnel**: While office personnel are commonly viewed as being somewhat "behind the scenes," they work tirelessly to ensure important communications are maintained, schedules are coordinated, meeting minutes are accurately generated, and similar administrative type duties are completed. In other words, these individuals ensure essential activities are completed to support day-to-day operations of the company.

Due to their routine interactions with senior and/or mid-management, they are usually more willing to accommodate new initiatives or changes to existing processes. These individuals are also usually more inclined to offer their insights while management is forming the implementation strategy. As such, they can serve as a good sounding board for informal brainstorming sessions.

○ **Project Personnel**: By definition, these individuals have a wide variety of roles and responsibilities. Examples include monitoring cost and schedules, leading routine project meetings, addressing personnel issues, spending time at the project location, and monitoring procurement of critical equipment and materials. Contingent upon their position within the project, they may also be responsible for meeting with clients to ensure expectations are met, as well as coordinating investigations of incident and/or injuries.

Due to the numerous duties and responsibilities already on their collective "plate," this group may prove to be somewhat hesitant to fully embrace your safety culture initiative. One approach to overcome this situation is to ensure benefits are also addressed, as well as options to reduce potential impact on their project while the initiative is being deployed.

○ **Support Organization Personnel**: For the purpose of this book, "Support Organization" addresses the wide variety of groups that assist with project execution. Examples include health and safety, quality assurance, procurement, human resources, and training. While these support organizations are critical to project success, I have encountered numerous situations where their contributions were not readily acknowledged.

On many occasions, the support organizations are not provided the opportunity to participate until the latter part of the project planning phase, or even worse, after the project has been kicked off. This inevitability results in cost overruns, schedule delays, last minute requests to regulators for special permits, etc. In turn, this can generate significant frustration within the support organization, which is commonly understaffed to begin with.

Due to their broad-based perspective, you need to ensure representatives from these support organizations are offered the opportunity to participate in your safety culture initiative as soon as possible. In the majority of instances, these personnel also routinely interact with task level personnel, which provides a great ally to assist with obtaining their buy-in.

○ **Task Level Personnel**: For the purposes of this book, task level personnel include technicians as well as craft in numerous disciplines.

3.0: Employees

Examples include: construction laborers, concrete masons, electricians, welders, ironworkers, teamsters, and janitorial staff. It is not uncommon for task level personnel to feel somewhat removed from senior management. This can be due to a series of factors. Examples include remoteness of work locations, night shift operations, or senior management having limited ability to be in the workplace on a routine basis.

Due to being somewhat "isolated," task level personnel can have a tendency to view senior management in a less than positive light. This can be inadvertently reinforced when these individuals were not afforded the opportunity to participate during development of previous initiatives. One unique aspect of this group is that they commonly do not readily "let go" of past incidents. In other words, they can hold a grudge. One approach is to personally meet with their superintendents, foremen, and task level personnel representatives to discuss frankly what is being proposed and solicit their assistance. By the way, this group can prove to be one of your strongest supporters, so it is definitely worth the effort to get their buy-in.

The balance of this chapter addresses numerous facets of employee engagement, including employee involvement. At the risk of stating the obvious, employees from throughout your organization need to be part of the solution.

__The employer generally gets the employees he deserves.__

– J. Paul Getty

3.1

EXAMPLES OF EMPLOYEE INVOLVEMENT

There is a wide variety of employee involvement processes, a few examples are provided below for your consideration. As part of your safety culture journey, you are encouraged to evaluate your current methods. Having said that, I also want to offer a few words of caution. Irrespective of the employee involvement process utilized, it is critical to ensure personnel are provided with feedback.

> **Irrespective of the employee involvement process utilized, it is critical to ensure personnel are provided with feedback.**

Suggestion Box

This was probably one of the first attempts used by many companies to stimulate employee involvement. This technique has small containers (with a slot in the lid to drop in the written suggestion) placed in company buildings. This method is fairly easy to initiate with minimal upfront costs. However, a fair amount of time can be consumed due to having to periodically check the boxes and retrieve the suggestion forms. Due to

3.1: Examples of Employee Involvement

competing priorities, what started out with a lot of energy can readily die on the vine for lack of ongoing management support.

I have visited many facilities that had some version of a suggestion box process in place. With the permission of management, I would randomly "sample" a few suggestion box locations by opening the lids and reviewing completed suggestion forms. On virtually every occasion, suggestion forms had remained in the box for months at a time versus being routinely retrieved and evaluated for potential implementation. In other instances, there were no suggestion forms placed in the boxes at all.

Some companies chose to make the suggestion boxes out of clear plastic so that it was easier to look in the boxes. While this approach met that goal, it inadvertently introduced another challenge when there were no suggestions forms visible in the box. In some instances, the suggestion forms had been recently removed by management for evaluation. Not realizing this, some employees perceive that the suggestion program is not being utilized. In other words, when employees see an empty suggestion box, they may think no one is submitting any suggestions, so why should they bother?

Informal discussions with employees at some companies noted their perception that the suggestion box process was not important to management. Examples included no response being provided back to the author of the suggestion, employees being rewarded for suggestions that could not be implemented, etc. Ironically, management at some of the same companies would proudly point out the suggestion boxes as one their "successes." Obviously, there was a significant communication gap between management and the employees when this scenario was encountered.

Good News: A suggestion box is easy to implement in the workplace.

Less Than Good News: This technique has a fair amount of "care and feeding" associated with it. Without ongoing emphasis, employees may not continue to use it.

Options: A wide variety of companies have transitioned over to an email-based suggestion program, which can increase ongoing participation. While the electronic suggestions still need to be evaluated, etc., this method significantly reduces the level of effort involved.

Safety Committees

Safety Committees are a relatively new phenomenon. The committee is comprised of representatives from the safety organization as well as personnel who perform routine work activities. These committees commonly have a sponsor from senior management and are tasked with looking at methods to enhance safety, assist with rollout of new safety initiatives, review of current processes, and similar activities.

While these committees can provide significant value, there is the very real potential for them to "go off the rails" and totally lose focus of their objectives. One of the companies that I worked with had two safety committees that had been in place for many years. One committee (Committee A) was very active throughout each year to enhance safety awareness. Examples included development of safety posters with their fellow craft members prominently featured in the pictures, performance of safety skits during monthly safety meetings, and using a 60 Minutes style interview format where craft would candidly discuss an injury they had in the past while working at other companies and what could have been done to prevent it.

3.1: Examples of Employee Involvement

Fairly recently, one of the Committee A members (Jimmy) had chosen to prepare a letter for his wife, explaining why he chose to die versus following safety rules. I had spent a lot time on the shop floor with this individual and appreciated how important his family was to him. I was also impressed that Jimmy had the courage to write such a letter and then read it aloud in front of his peers during a routine safety meeting.

He had to pause numerous times while reading the letter in an attempt to keep his emotions in check. This became readily evident when he talked about his brother-in-law having to take his sons to softball practice. By the time he finished, there wasn't a dry eye in the house, mine included.

After the meeting, many of his team members approached Jimmy, thanking him for sharing the letter and the reasons behind it. What was really intriguing was how many of his peers committed to writing their own letters, which they then shared during subsequent meetings. By the way, the emotions generated while listening to Jimmy read the letter still resonate with me to this day.

In subsequent discussions with Jimmy at his work location, I learned that while he hadn't shared the letter with his wife, he reads it almost weekly during breaks at work. He found that the simple act of writing the letter reinforced how important his family is to him and, in turn, how following safe work requirements allows him to go home safe each day.

By way of comparison, the other safety committee (Committee B) did very little, other than meet once a month and share their concerns with management. While having a venue to communicate concerns is advantageous, the committee had degraded to the point that this became the main focus of the meeting. Even more disturbing, the chair of the committee was left to her own devices when attempting to generate topics for the meetings, identifying potential safety improvements, etc.

One method the company utilized to enhance awareness of safety was to select personnel to attend a large safety conference each year. While a lot of individuals wanted to attend the conference, the actual number of attendees was contingent upon the funding that was available for a given year as well as current mission tempo.

Due to a reduction in funding one year, the company wasn't able to send as many people as in the past. While members of Committee A fully understood the situation, members of Committee B became very angry and complained all the way up to the president of the company. The president asked me to look into the process, which generated some interesting results. Perhaps the most notable was the company did not have any method to determine who should attend. Without such a process, some personnel (as in Committee B members) had come to view attendance at the safety conference as a "right" versus a "privilege."

To assist with enhancing safety awareness and commitment to the company's continuous improvement goals, I worked with representatives of both committees to develop a series of criteria that had to be met prior to being considered to attend the annual safety conference. Unsurprisingly, there were significantly more members from Committee A who assisted with development of the criteria versus members from Committee B. Some members of Committee B actually decided to quit their committee since they weren't "guaranteed" the right to attend the conference.

Good News: Safety Committees can prove to be a great resource to enhance safety processes.

3.1: Examples of Employee Involvement

Less Than Good News: Without clear roles and responsibilities, the committee can't function to its maximum capability and may ultimately lose its focus.

Options: Ensure the safety committee has a comprehensive charter as well as ongoing support from the management team. Some companies will have a senior member of the safety organization serve on the committee who can assist with establishing goals, initiatives, and similar techniques to help keep safety "fresh." Routing of meeting minutes addressing safety committee discussions to the senior management team can also assist with monitoring committee efforts.

Integrated Project Planning Teams

Numerous companies have established project planning teams that are comprised of personnel from groups across the company. These integrated teams are engaged early in the project planning phase to assist with identifying topics that need to be addressed prior to and during the work evolution. Examples include procurement of equipment and materials, special permits required for authorization of work, and staffing considerations.

While it would appear that the advantages of this technique should be self-explanatory, some companies have not chosen to embrace this technique. The justifications include not having enough time for the planning meeting or the expense of holding the meetings. Ironically, these same companies commonly encounter cost overruns, schedule delays, or rework due to not involving the correct people early in the planning phase.

By the way, I'm old enough to recall a Fram oil filter commercial with the tag line "*Pay me now or pay me later.*" The message that was being conveyed in that television commercial was that proper scheduling of routine maintenance can save the engine in your car. The same strategy can be said about saving a project, activity, or initiative.

This attitude of not needing to address integrated project planning has been encountered in both small and large companies. I have worked with organizations with annual contract values approaching $500,000,000 per fiscal year, with hundreds, and in some instances, thousands of personnel. In other words, while these organizations, by definition, had the capability to incorporate integrated project planning teams, their "culture" had not truly embraced the concept.

Good News: Integrated Project Planning Teams can provide significant value during the project planning phase and subsequent work evolution.

Less Than Good News: Senior management buy-in is essential to overcome potential resistance at the mid-management and/or lower tier levels.

Options: In the majority of instances, most companies that do not utilize integrated project planning teams have experienced numerous challenges, including rework or schedule delays. In turn, these examples can illustrate the value that could be gained from use of these teams. Some of the best examples can be provided by the personnel actually responsible for performance of the work. In turn, this information can be utilized while conducting a "pilot" exercise (e.g., a project of relatively smaller scope) to prove the viability of the concept prior to full deployment.

3.1: Examples of Employee Involvement

Employee Surveys

In the current work climate, most companies have some version of an employee survey. These can include hard copy reports, working groups, or email polls. Regardless of the technique selected, a focused "campaign" needs to be utilized whereby the overall strategy is shared throughout the organization. This campaign should also address how the survey results will be utilized.

I have worked with companies that dedicated the better part of a year focusing on the campaign strategy prior to actually routing the survey to employees. This approach directly attributed to results that proved beneficial to the company. Examples included improved communications, identification of potential process improvements, in addition to further enhancing their overall safety culture.

Similar to earlier discussions, this approach should be self-evident for any company. Unfortunately, over the years, I have also encountered companies that dedicated little, if any, level of effort prior to implementing the survey. In some instances, employees only received an email announcing the survey, with the actual survey being issued for their use shortly thereafter. The resulting survey results provided significant challenges.

Examples included personnel not understanding the questions, "checking the boxes" to quickly complete the task, or simply ignoring the request to complete the survey. While the collective results did not actually reflect the overall health of the company, mid-level managers may be still required to develop "get well" plans addressing survey results that were less than desirable. The end result can be a missed opportunity to address overall employee satisfaction in addition to generating more frustration between employees and the management team.

Good News: Employee surveys can provide an accurate portrayal of employee satisfaction as well as the capability to compare results from prior employee surveys.

Less Than Good News: Without adequate effective up-front planning, including an associated communication campaign, the results might not accurately reflect actual conditions and may do more harm than good.

Options: Prior to beginning the campaign, ensure your senior management team is aware of the need for an effective communications strategy. Also ensure the team appreciates and supports the need to allow some "soak time" so that employees understand what is expected of them when participating in the survey. The soak time phase can also be utilized to provide additional clarity and/or have small groups brainstorm options to embrace the survey.

Use of a "Buddy System"

Regardless of the company, new employees are faced with a series of challenges when entering the work place, including having to adjust to new "norms" as well as fitting into the organization. While all companies have some version of a new hire indoctrination program, an additional technique that is gaining traction is the use of a "buddy system." This process aligns the new employee with a current member of the workforce, preferably someone within the work group/team of the new employee. This technique has provided numerous advantages, including reducing the learning curve regarding use of company procedures, and more

3.1: Examples of Employee Involvement

importantly, enhancing the new employee's ability to feel that they are part of the company.

Very recently, my wife's great-nephew (Patrick) was thrilled about being hired for a new job with a small local television broadcaster located in Northern California. While this required Patrick to move across the country, he was looking forward to beginning a new career. He was in his late twenties, and like a lot of us at that point in our lives, was convinced he could readily succeed. With all of his possessions loaded into his small import automobile, Patrick drove for three days and then began settling into his new apartment.

After meeting with his new boss, he received a series of local news assignments to cover. The assignments consisted of Patrick having to develop the story line, conduct filming and interviews on location, and then compile the completed digital files and send them back to the studio. While this could be viewed as a dream job, Patrick had little, if any, support from his new boss or other members of the team. Candidly, he was left to work on his own almost every day, having almost no interaction with other personnel at the studio. This situation became so frustrating for Patrick that he finally resigned and drove back across the country to temporarily move back in with his parents.

Although that position didn't work out, Patrick continued to pursue his dream of broadcasting. While reviewing a series of job offers, he received a call from the same local broadcast affiliate in Northern California, asking him to come back to work for them! Turns out they really liked the quality of his work and also appreciated the need to have a process for new employees to be partnered with a current member of the team. They also shared with Patrick that he would be invited to attend routine meetings with the broadcast team as well as have the opportunity to attend a series of conferences and networking sessions so that he could expand his knowledge and skill set.

After talking with his parents as well as Mary and me, Patrick decided to return to the company in Northern California. Patrick now really enjoys having a work partner at the studio to help him understand new processes and fit into the company. Due to the changes in the work environment, Patrick's attitude was amazing. He looks forward to being at work each day and truly feels that he is a member of the team versus just someone who is stuck "in the field" all the time. By the way, even though this company has less than fifteen people, the need for having a consistent methodology to help new employees become part of the team proved just as essential as for a large company.

Good News: Use of a Buddy System can help with integrating new personnel as well as jumpstart their understanding of the company safety culture.

Less Than Good News: By definition, there is some impact on the existing employee who agrees to serve as the buddy. Examples include completion of routine work assignments, participation in meetings, and similar activities.

Options: To reduce impact on the existing employee serving as the buddy, examine options whereby others can assist with routine work assignments. In some instances, the new employee can accompany the buddy to a scheduled meeting. This approach allows the buddy to meet an obligation while also providing the new employee the opportunity to experience a meeting. Another consideration is to establish a "pool" of current employees wishing to serve as buddies for new employees so that the buddy assignment can be rotated.

Volunteerism

Being part of the local community has proven on numerous occasions to strengthen a company's ability to attract talent and reduce attrition. The majority of companies have chosen to weave "Social Responsibility" into their organizational construct, including their core values. Examples of social responsibility vary, including serving on the board of directors for local charities, establishing funding for charitable organizations meeting selection criteria, and encouraging employees to serve as speakers during career days at local schools.

Another approach to embracing social responsibility that is gaining more acceptance is providing opportunities for employees to support local charitable organizations by serving as volunteers. Examples can range from assisting with soup kitchens to serving on the planning committee for a favorite charity. Some companies provide financial support to charities based on the number of charitable hours worked by the employee. Employees can also be allocated a designated number of hours each month to assist a charity. The benefits from any of these approaches are numerous, including providing critical support for charities in need and enhancing the quality of life for the employee.

Good News: Companies supporting local charities commonly realize significant value from the expenditures. Employees participating in the volunteer efforts gain a "big picture" appreciation regarding how the company and, in turn, themselves are making a difference to those in need.

Less Than Good News: Regardless of the option selected, supporting volunteer efforts requires some expenditure of company funds.

Options: To reduce unanticipated impact on company funds, ensure this activity is addressed during development of annual budget forecasts. As an aside, many companies also address their commitment to support local charities in their marketing campaigns, which can provide another ancillary benefit.

President's Safety Council

Separate from Employee Safety Committees, establishing a President's Safety Council can further enhance employee involvement as well as communication. This type of committee is commonly led by the most senior level of management, with attendance by employees from throughout the company. Taking into account scheduling commitments for senior team members, this type of meeting is usually held on a quarterly basis. To provide maximum value, the meeting is utilized as a forum whereby both senior management and task level personnel share experiences in the workplace. Employees in attendance can also share concerns or topics that need to be addressed. However, it is critical that this meeting does not circumvent established chain of command processes that are in place for use by employees and their line management.

This type of meeting has proven to further enhance relationships between the management team and employees throughout the company. This strategy can provide significant value for large organizations where routine face-to-face interaction with employees has proven to be a challenge.

3.1: Examples of Employee Involvement

Good News: This type of meeting can greatly assist enhanced employee communication as well as build increased rapport between senior management and task level personnel.

Less Than Good News: Senior management must be willing to commit to attending the meetings as well as not inadvertently overriding mid-level management decisions. Another challenge that can be encountered is personnel not being willing to initially discuss topics in an open and candid manner.

Options: Use of a charter can assist with clarifying the purpose of the meeting as well as anticipated outcomes. Designating an individual with demonstrated expertise as a facilitator can assist with having an open forum whereby attendees feel comfortable becoming engaged.

By the way, one of the common themes you'll note throughout this book is employee engagement/involvement. This is by design, since being able to gain employee participation and acceptance needs to be one of the key steps in your safety culture journey.

He who is not courageous enough to take risks will accomplish nothing in life.

— Muhammad Ali

3.2

Respect in the Workplace

One topic that is often overlooked is ensuring personnel are treated with respect. Having respect within the company, especially at the task level, is a foundational element to support enhancing the overall safety culture of any

> One topic that is often overlooked is ensuring personnel are treated with respect.

company. Ideally, the company aligns this commitment with the broader construct of tenets at the corporate level. Unsurprisingly, there have been numerous articles written and publications issued addressing the importance of respect in the workplace.

As discussed in Chapter 1, the INPO organization focuses on the commercial nuclear power industry and has issued a series of publications addressing safety culture. One of my personal favorites is the approach provided in INPO 09-011, "*Achieving Excellence in Performance Improvement, Leader and Individual Behaviors that Exemplify Problem Prevention, Detection, and Correction as a Shared Value and Core Business Practice*" (Institute of Nuclear Power Operations, 2009). What intrigues me with this publication is the approach taken whereby company performance is aligned with the behavior expectations of individuals throughout the company.

This publication addresses a series of desired "outcomes" regarding performance improvement at the company level. Examples include the picture of excellence is well known, problems are prevented, mistakes are avoided, and performance improvement is ingrained as a core business value. There are also a series of behavioral aspects provided that support implementation of the outcomes. To provide maximum value, different behaviors are identified at the individual, line manager (e.g., supervisors), and senior management levels.

Behavioral examples at the individual level include reporting problems to supervisors, applying learning gained from training into daily work activities, and embracing performance improvement actions. Behavioral examples at the line management level include understanding how work is really performed, being receptive to reports of incidents and near misses, and using appropriate performance metrics. Finally, behavioral examples at the senior management level include establishing challenging performance goals, promoting a vision to detect/prevent problems, and rewarding personnel who identify and/or resolve issues.

I have successfully utilized these concepts to develop similar criteria for a series of companies. This approach provides numerous benefits, including tailoring the criteria to address company specific expectations and tenets. In a similar manner, binning performance expectations with the company criteria illustrates why the desired performance expectations are important. To gain the maximum traction (e.g., sustainability), employees from throughout the company were provided the opportunity to participate in development of the criteria and expectations.

The resulting criteria can directly contribute to enhancing respect not only at the task level, but throughout the company. I have utilized these criteria during staff sessions, plan of the day meetings, and similar gatherings, to lead discussions regarding performance expectations for the different levels (e.g., task level versus line management versus senior management). On numerous occasions, these discussions resulted in task level personnel gaining additional perspective regarding the difficult role served by senior management. In a similar manner, senior level management gained additional insight regarding performance expectations across the company. These criteria can also be utilized during investigations of near miss events, incidents, and injuries to gain insight regarding company programs and/or processes that need to be revised or enhanced.

As part of this collective effort, there is an additional criterion that should also be included in any and all endeavors addressing respect in the workplace. Namely: Be your brother's keeper. Note: In recognition of the expanded role of women in the workplace, this phrase has now been expanded to read: Be your brother's/sister's keeper. For ease of use in this book, I'm using your *brother's* keeper.

> ... Be your brother's keeper.

This phrase has been addressed by the safety community in numerous venues, and I have included this topic while serving as a speaker at safety conferences. The premise of being your brother's keeper is that few, if any, workers truly perform work on their own. The majority of companies have implemented procedures to follow when performing more hazardous tasks (e.g., working at elevation, conducting lockout/tagout of energized sources, working at remote locations). In most instances, at least two personnel are assigned to perform the task together. For less hazardous work activities that may be conducted by an individual worker, that person would still participate in some type of meeting to discuss the work to be performed that day.

By being your brother's keeper, personnel embrace the concept that accidents can be prevented and having an additional set of eyes can assist with safe and compliant performance of work. I am intentionally emphasizing the need for task level personnel to embrace this concept. In other words, they need to own it.

Unlike some safety culture concepts that are a little harder to discern, successfully implementing being your brother's keeper in the workplace can be readily evident. Personnel intentionally monitor for potential hazards that other personnel may be exposed to during performance of their work and are aware of events that could occur. In addition to Plan of the Day/Tailgate Safety Meetings, the topic of potential hazards is also routinely shared among work teams. These same individuals do not permit anyone to take short cuts and are receptive to feedback being provided to them by their peers. When you have inculcated the concept of being your brother's keeper into your work environment, you are well on your way to a sustainable safety culture journey. I highly encourage you to include this topic in your company processes.

Before leaving the topic regarding respect in the workplace, I wanted to share a fairly recent example whereby a well-intentioned gathering between craft personnel and their leader inadvertently came completely off the rails. The premise was a good one: include free breakfast during the traditional morning meeting where craft talked about work to be conducted that day, hosted by a leader from the company. To provide a conducive meeting environment, tables and chairs were placed in a U shape configuration, with the leader sitting in the middle between the long legs of the "U."

I was invited to sit in on the meeting, which had approximately 25 craft personnel in attendance. All attendees, including their foremen and superintendents, ultimately reported to the leader who hosted the meeting. By the way, this leader (Richard) had a history of being perceived as a bully who also overreacted to work place incidents. On numerous occasions, he was observed actually screaming at the individuals performing the work when he didn't like what he saw. This situation was compounded due to peers being in the immediate work area when the chastising occurred.

Richard kicked off the breakfast meeting with thanking everyone for their attendance and then listened to the discussions regarding work activities scheduled for that day. He then offered his personal thoughts regarding safety in the workplace and asked for everyone in attendance to provide their insights. Even though some of the craft didn't appear comfortable sharing their opinions, Richard continued to push the topic. This began to generate some friction since some individuals were quieter than others and/or didn't trust the leader. Unfortunately, Richard didn't appreciate some of the personality types of the craft since he hadn't spent much time with them in the workplace.

After a period of awkward silence, one of the craft (Billy) finally spoke up and began to discuss lack of new equipment that the group needed. It was very apparent that this topic was important to Billy, and he provided a series of examples where new equipment could help their day-to-day activities. By the way, Billy was sitting immediately to the left of Richard.

As soon as Billy began sharing his thoughts, Richard proceeded to eat his breakfast and drink his coffee, while looking to his immediate right. Again, Billy was sitting on his left-hand side, but Richard continually looked to his right or simply stared up at the ceiling! I observed this interaction, including noting that Billy was quickly becoming frustrated.

3.2: Respect in the Workplace

After Billy finished speaking, Richard thanked him for his thoughts, promising to look into them. The breakfast meeting concluded fairly quickly thereafter.

I was standing beside Richard as people were beginning to leave, and after Billy stood up, he turned to Richard, stating that he had one more question. After Richard told him to go ahead, Billy looked Richard squarely in the eye and simply asked "What's my name?" I was dumbfounded when it became readily apparent that Richard couldn't recall the name of an individual who had sat beside him for the better part of an hour and shared concerns that he felt were important. To compound the magnitude of this management gaffe, Billy's name was embroidered on the front of his crew shirt! Billy then stated, "By the way, my name is Billy," and then walked away with some of his fellow workers.

While the meeting ended on a fairly awkward note, the situation did not improve during a subsequent meeting I had with Richard. He was very upset that Billy did not treat him with "respect," and it became clear that he felt the meeting was a total waste of time. In other words, Richard completely missed a great opportunity to begin reframing his relationship with his personnel.

This situation was compounded by Billy's feelings of resentment towards Richard, which I found totally understandable. After all, Richard appeared to ignore him and then couldn't even remember his name. Having said that, Billy unfortunately chose to hold a grudge toward Richard and shared his feelings with other craft personnel. The animosity associated with this event was so visceral that it remained a topic of discussion with the craft group over a year later.

No one, including myself, envisioned the meeting devolving so quickly, and the negative effects proved to be long lasting. While Billy could have handled the scenario differently, I was most disappointed with how Richard conducted himself during the meeting. To coin an old phrase, respect is something that is earned, not something automatically given (Lee G. Bolman & Terrence E. Deal, 2003). Based on this litmus test, Richard failed on all counts. Despite this interaction, Billy chose to maintain a high degree of pride and ownership when performing his work. The next subchapter provides additional insight regarding this topic.

Great leaders have a heart for people.
They take time for people. They view people
as the bottom line, not as a tool to get to the bottom line.

– Pat Williams

3.3

Pride and Ownership

When addressing safety culture, the topic of personnel taking pride in their work and willingness to own the final product also needs to be on the forefront. By way of comparison, think about instances where you've had to select a new automotive repair shop to work on your vehicle.

> When addressing safety culture, the topic of personnel taking pride in their work and willingness to own the final product also needs to be on the forefront.

Unless you have a great referral from a friend or co-worker, if you're like most of us, you're going to check out a series of repair shops before finalizing your selection. While you probably won't be carrying around a detailed checklist, you're undoubtedly mentally observing a series of conditions.

Examples could include overall cleanliness of the service bays, tools being appropriately stored in toolboxes, the waiting area for customers is clean and has comfortable seating, and the repairmen are appropriately dressed. In addition to the overall condition of the repair shop, your decision is also going to be influenced by your interactions with the owner.

3.3: Pride and Ownership

Topics can include: Was he receptive to your concerns/needs? How does he conduct himself with his employees? And how well he is interacting with other customers?

Many years ago, I was in the waiting area of a repair shop located in Omaha, Nebraska. I wanted to speak with the owner about some routine maintenance that needed to be performed on my automobile. I was stunned to observe the owner screaming at his employees, telling them how stupid they were for not getting the work done correctly and on time. He also made some disparaging remarks about former customers. I could tell that the receptionist was very embarrassed by his outburst, and she quietly apologized to me after he left the waiting room, also mentioning that she had witnessed this behavior before. Needless to say, I did not feel comfortable having his shop work on my car. By the way, I made this decision without having any insight regarding the overall quality of their actual work. In other words, his horrific approach to treating his employees and customer relations overshadowed the actual technical capabilities of his shop to service/repair vehicles.

Shifting from the automotive repair shop scenario to your company, one of the ways I've found to get a sense of the level of pride in the workplace is to occasionally walk through break rooms, kitchen areas, equipment rooms, and repair shops. Hopefully, you will observe equipment that is clean, tools that are appropriately stowed, and break rooms/kitchen areas that demonstrate good housekeeping. Conversely, if you encounter equipment/tools that are not clean/scattered about or kitchen areas with old/moldy foodstuffs laying on tables and counter surfaces, your personnel are not taking ownership of their areas.

While recently working with a large company, I noted that a commercial size coffee maker in one kitchen area was not being consistently turned off at the end of the day. This type of coffee maker has a series of heating elements to ensure the coffee remains hot, with the capacity to hold up to four carafes of coffee on the heating elements. I was usually one of the first to arrive at the work location each morning. At least 4–6 times a month, I noted the aroma of burnt coffee when entering the area. The odor was due to the heating elements of the coffee maker being left on overnight, thereby burning off the residual coffee that had been left in the carafes. In some instances, the coffee maker was left on over the weekend while no one was in the office.

On a series of occasions, I spoke with the managers responsible for the area regarding the potential hazards associated with this practice. Examples included being exposed to broken glass shards and thermal burns while attempting to handle and clean the hot carafes. Despite repeated requests from their management, no one in the immediate work area would take responsibility to ensure the coffee maker was turned off at the end of the shift.

This group (Group A) of approximately thirty personnel also had access to a total of three refrigerators. As with most kitchen areas, the refrigerator doors had clear postings advising users to not leave foodstuffs overnight and stating that the units would be cleaned out each Friday afternoon. Unsurprisingly, there were numerous "science experiments" growing in the refrigerators. It was not uncommon for items to be left in the refrigerators for weeks at a time. There was also various debris, old containers, foodstuffs, dirty appliances (e.g., microwave ovens), and out of date flyers scattered about on the kitchen counters and seating areas.

When challenged about the conditions, personnel who worked in the immediate area had a series of common responses/excuses. Examples included "Hey, it's not my job," "Must be someone else's trash," "I don't drink coffee," etc. In other words, they refused to take any ownership of the issue and were more than willing to blame someone else.

3.3: Pride and Ownership

I commonly refer to this as the SODDI defense: Some Other Dude Did It. What is most concerning with the SODDI defense is how it can migrate into routine work activities, including associated deliverables. This has proven very problematic for many companies, especially those who do not have consistent disciplinary action processes in place.

> I commonly refer to this as the SODDI defense: Some Other Dude Did It.

Middle management finally got so fed up with the situation that they had the coffee maker and refrigerators removed. Well … you can readily predict what happened next. Yep, the personnel working in the immediate area complained to anyone who would listen about what had been taken away from them. Some of them even went so far as to submit a complaint to the Human Resources Department, convinced their "rights" had been violated. This situation continued to escalate, including some individuals actually being issued written warnings for insubordination.

At the other end of the spectrum, I observed another kitchen area that had a similar configuration, including a large coffee maker with multiple carafes and two refrigerators. I am embarrassed to admit that their kitchen area was probably cleaner than the one at my house when my wife is out of town. I had the opportunity to speak with some of the personnel who used that area, and their "ownership" was very clear. They could readily explain the importance of having someone assigned each week to make sure the coffee maker was turned off each night as well as reminding their co-workers to remove their containers, etc., from the refrigerators at the end of the week. They even took turns cleaning the counter spaces, sinks, and appliances each Friday afternoon. That group (Group B) definitely demonstrated pride in their workplace, and this same pride was also reflected in their work activities, including project deliverables.

Here's the ultimate irony. Not only were these two groups part of the same company, they were located in the same building. Not only were they located in the same building, they were on the same floor, approximately one hundred feet down the hall from each other! While Group B was in close proximity to Group A geographically, their approach to maintaining their work environment, their deliverables, and their overall "culture" was worlds apart.

Further investigation determined that the two groups reported to different managers. The managers of Group A were very hands off in their approaches to managing their personnel. Refer to Chapter 2 for additional insight regarding laissez-faire leadership style. They were more focused on their own careers, routinely cancelling group meetings, working in their offices with the doors shut, etc. Their approach to management directly contributed to an enabling atmosphere where personnel were not being held accountable for their actions. In turn, numerous personnel were under the impression that they didn't need to care about the work environment. The condition of their individual workspaces (e.g., cubicles) was very similar to those encountered in their kitchen area. Namely, excessive clutter, combined with numerous tripping hazards, poor posture presenting potential ergonomic issues, etc. Unsurprisingly, project deliverables from Group A also had a fairly high rejection rate and/or due dates being missed.

The manager for Group B had taken a completely different approach. The topic of maintaining their workspaces in a professional manner was routinely discussed during staff meetings. If the manager had a conflict, he ensured the meeting was still held with a designee formally designated to run the meeting. This approach also assisted with providing growth opportunities for individuals throughout his group. Regarding maintaining the kitchen area and coffee maker, the group had developed their

3.3: Pride and Ownership

own schedule and assigned the tasks on a rotating basis. The manager actually developed a friendly competition where "awards" were handed out during the staff meetings for overall cleanliness, etc. In other words, the manager of Group B worked in a collaborative manner with his team to address the issue and develop solutions. This manager utilized a similar approach to addressing project tasks and development of deliverables, including methods for the group to check each other's work. Referring once again to Chapter 2, the Group B manager exhibited numerous traits associated with having a democratic leadership style.

After reviewing the results of my investigation, senior management made a series of changes, including replacing the managers of Group A. While not initially removed from the company, the managers were demoted and placed on detailed improvement plans to determine their ability to be retained by the company. Group A personnel also had to meet with Human Resources representatives and develop similar improvement plans. While some members of the group supported the process, others chose to blame the prior managers for any problems and refused to change their attitudes, including their work ethic. In other words, they chose to continue their SODDI attitude. Needless to say, a series of individuals from Group A are no longer with the company.

Conversely, the manager of Group B was asked to share his approach with the senior management team. While he graciously accepted the meeting request, he also ensured that some of the members of his group were invited to be at the meeting. He kicked off the meeting by introducing his team and then turned the presentation over to them. The presenters really enjoyed the opportunity to meet with the senior team, and numerous members of Group B subsequently served as mentors to other groups who wanted to initiate similar processes. This is also a great example of how combining effective leadership with opportunities for employee engagement provides numerous benefits.

As a further illustration regarding pride and ownership, I recall a story about three individuals who were performing the same job. Namely, laying bricks. When asked by a local newspaper reporter what their jobs were, the responses were very insightful. The first worker responded in a fairly heated manner, "Can't you see I am laying bricks?" The second worker sighed and stated, "My job is to build this stupid wall." The third worker looked at the reporter, and with a smile on his face and in his voice, responded, "I am helping construct the greatest building in the city!"

Hopefully, you appreciate how the different attitudes demonstrated by the bricklayers can directly affect their work. The first worker was grudgingly going through the motions, while the second felt that at least he had a job. Conversely, the third worker demonstrated that he had a clear picture regarding how his contributions were making a difference. In other words, he could see the big picture. If given the choice, I will pick the third bricklayer to build a new wall for me every time. Regardless of the position or assignment, I am convinced that the day-to-day work activities of any individual can be enhanced by taking pride in their work.

I am also a proud graduate of the Ziglar Institute located in Dallas, Texas. This institute was founded to share the teachings of Zig Ziglar. In his book entitled *Born to Win, Find Your Success Code*, Zig discussed taking pride in your work and being the best you can be, regardless of the task being performed (Zig Ziglar, Tom Ziglar, 2012). One of my favorite Zig stories is how your attitude can directly impact your job and career.

His story focuses on a flagman who was part of a road construction crew. This individual served an important role regarding safe performance

3.3: Pride and Ownership

of the work by monitoring when vehicles needed to be stopped so that construction equipment operators could perform different tasks. The majority of us have encountered them when we are driving. On numerous occasions, I've observed the flagman simply going through the motions. In some cases, they were barely paying any attention to the activities occurring around them while checking their cellphone or daydreaming.

Compare those individuals to the flagman that Zig discusses. This flagman goes out of his way to be courteous to every driver passing by. He practices in front of a mirror to observe his posture and makes sure that he is wearing clean work clothes each day. Instead of slouching against a company vehicle and glaring at people as they pass by, he is standing upright and smiling.

The flagman also concentrates on waving his flag in a professional manner and always ensures he has a smile on his face. In addition to clearly demonstrating pride in his work, his efforts are going to make a great impression on other members of the construction crew as well as those driving by. Unsurprisingly, one of the people who was impressed was his boss. This positive impression will probably lead to a promotion or more challenging job assignments, simply because the flagman made an intentional decision to take pride in his work.

I am old enough to recall a television commercial with a tag line of "*be proud enough of your work to sign it.*" The commercial closed with a worker at an automotive assembly line signing his name with spray paint on the inside of a front fender that he was placing on the new car. In summary, if your personnel do not have pride in their workplace, how can you reasonably expect them to take pride in their deliverables?

By the way, I'll fully understand if you decide to momentarily put down this book and go check the break/kitchen areas at your office, plant, and other locations. Hopefully, you will be pleased with the results of your surprise tour. If not, you've identified another topic for your safety culture journey as well as a potential metric that could be included for monitoring long-term implementation.

The next chapter provides some insight regarding options to consider as you begin your safety culture journey. You are encouraged to pay particular attention to the discussions addressing use of a team approach as well as options to address challenges that will undoubtedly be encountered during your efforts.

If a man/woman is called to be street sweeper…
he/she should sweep streets so well that all the hosts
of heaven and earth will pause to say, here lived
a great street sweeper who did his/her job well.

– Martin Luther King, Jr.

4.0

WHERE DO I START?

As has been said many times, "Begin with the beginning." In the case of safety culture, this entails a series of key actions to help start your journey. You're undoubtedly noticing that I use the word *journey* on numerous occasions in this book. This is because this effort needs to be viewed as an ongoing effort versus simply issuing some procedures, hanging a few posters, and then attempting to declare success.

Prior to initiating an endeavor of this magnitude, you need to spend time determining what is driving the need. Examples could include increased injuries, excessive attrition, customer complaints, extensive rework of products, or perhaps just an uncomfortable feeling that something is just "not right." Another way to look at this is: "What keeps you awake at night?" When you start to experience concerns beyond those commonly associated with maintaining market share, that can be an indication that the culture of your organization could be enhanced.

> Another way to look at this is: "What keeps you awake at night?"

4.0: Where Do I Start?

That type of feeling led to an initial meeting I had with the owners of a large manufacturing company. Over the prior few months, they had been getting more concerned about the potential for someone to be permanently injured. In other words, they thought their employees were not working safely. In this instance, the owners did not really have definitive information, just an uneasy feeling (e.g., something was keeping them awake at night). During a series of subsequent meetings and spending time on the floor with their employees, it was determined that senior management did not routinely engage with personnel performing the day-to-day work. Since the employees rarely saw the owners in the workplace, the employees had a perception that the "big bosses" didn't care about their safety. On the rare occasion when the owners attempted to speak with the employees, there was little engagement, so the employees simply stated, "Everything's fine." This contributed to the owners being uncomfortable, nervous, and staying "awake at night."

In this case, I was able to determine that personnel were performing work in a safe manner in the majority of instances. However, it was also determined that there was not an effective communications process in place between senior management and the task level personnel. Ironically, this situation also stifled potential process improvements addressing performance of daily work activities.

With the support of the management team and the workers, a series of processes were rolled out over the next few months. Topics included enhanced management engagement with the workforce as well as having a formal method to capture potential process improvements for work practices. Awards presented for process improvements were also highly publicized in the company newsletter as well as being discussed during routine staff meetings.

The end result dramatically shifted their safety culture and provided the owners with one less topic to keep them awake at night. Perhaps the most significant achievement for this company was the enhanced relationship with their task level personnel. As the process improvements gained traction, task level personnel really looked forward to having the management team in the workplace. There was also a reduction in unsafe work practices. Collectively, the organization became more of a family versus an "us versus them" mentality.

Regardless of your reasons, you are to be congratulated on having the courage to self-reflect and conclude that you want to enhance your safety culture. I also encourage you to consider working with those who have expertise in this area. They can bring an independent perspective and a fresh set of eyes to assist your efforts.

Now a word of caution, there are numerous companies in the industry more than willing to sell you the "perfect fix" to address your cultural challenges. By way of example, one company recommends raffling a new pickup truck, with the drawing commonly being held during the month of December. However, individuals who got injured during the year would be excluded from being able to win the pickup. While well intentioned, there have been many instances where use of this raffle process resulted in driving (pardon the pun) reports of injuries underground.

In other words, personnel are still getting injured; they simply are not reporting the injuries. This process can also generate frustration when other workers are aware that the winner of the pickup had gotten injured that year but didn't formally report it. More importantly, personnel may experience long-term manifestations due to unreported injuries not being properly treated by a physician.

4.0: Where Do I Start?

This scenario can result in your company having the potential for personnel not being able to work, in addition to impacting their quality of life due to lost wages. You also have expended a significant portion of your budget allocated for safety improvements that may actually reduce the effectiveness of your overall program.

This is one of a myriad of examples available to unsuspecting companies. Most susceptible are those companies looking for a quick fix, or perhaps a solution that doesn't require a lot of effort on their part. In my opinion, being able to achieve successful implementation of an effective safety culture program is the culmination of a focused, comprehensive, concerted effort that is ongoing.

> ... Fast, Good, or Cheap
> ... pick two!

By the way, when I was growing up, I spent a lot of summers working on my uncle's farm in northern Missouri. I learned a lot from him over the years, including the ability to laugh at myself and have a strong work ethic. Although he probably had no more than a fifth grade education, he was one of the smartest men I ever met, especially when it came to common sense. One of his favorite phrases was "Fast, Good, or Cheap … pick two!"

I rather like that analogy when viewing a safety culture journey. In other words, avoid sacrificing sound approaches to grab an off the shelf solution. In a similar manner, the cheapest software available via a "Buy It Now" sales pitch offered by a vendor at a safety conference may not be worth the long-term cost. Regardless of the discount being offered, the software will undoubtedly still have to be tailored to meet the needs of your company.

Ideally, you are able to reach out to owners of similar companies to gain some insight regarding the processes they have used. I also recommend that you screen any consultants being considered. In addition to checking references, you can review technical papers they have written or books they have published. This overall approach can help you determine if their philosophy and approach align with the tenets of your company.

If you have the opportunity, I also highly recommend that you also consider attending safety conferences in your area. The majority of these conferences have an exhibitor area where you can meet a wide variety of safety consultants. Again, you want to select a consultant that you are comfortable working with. Hopefully you'll be working together for a long time!

During the initial planning phase, you are encouraged to gather your senior team to discuss the initiative. Ideally, this meeting would be held away from the office to reduce potential distractions was well as demonstrate that this topic is something other than routine.

At this meeting, you want to be open and candid regarding why everyone is gathered together. You could also offer some examples of "ideal" safety cultures at other companies, and then ask your team for their perspectives. Refer to Chapter 1.3 for additional insight. To help "prime the pump" so to speak, you can share these thoughts in the meeting invitation, and ask them to be prepared to share their thoughts. I have found this approach also helps some attendees who don't like to speak without preparing their thoughts in advance.

Because many people aren't necessarily comfortable with voicing their opinions, this approach may feel a little awkward at first. Having someone facilitate the meeting for you can assist with the process and help ensure that everyone's thoughts are addressed. I have facilitated numerous sessions for working groups ranging from a handful of attendees to well over one hundred, and historically found that this strategy provides a series of benefits. Examples include having a "neutral" voice in the room, being able to defer topics that are not directly germane to the discussion, and ensuring all participants have an opportunity to share their thoughts. An effective facilitator can also assist with determining next steps to maintain momentum versus simply closing out the meeting without any definitive calls to action.

During this initial phase, I encourage you to avoid becoming overly focused on having a "perfect" planning meeting. By allowing the process to have its natural ebb and flow, participants will appreciate the ability to revise the overall effort as the strategy is being finalized. This is another advantage of utilizing a facilitator since they can guide the overall effort without inadvertently interfering with the creative process.

I also highly recommend that you initially withhold your thoughts until others have spoken. This usually provides for more of an open free flowing forum versus some people feeling intimidated since "the boss" has already stated what needs to be done. The balance of this chapter addresses a series of topics, including the value provided from utilizing a team approach, selection of champions, and practicing deliberate speed.

The secret of getting ahead is getting started.

– Mark Twain

4.1

The Need for a Team Approach

As with the majority of successful efforts, the importance of utilizing a team approach cannot be overemphasized. When I use the word *team*, I am referring to much more than just a small group of individuals from your safety organization. Ideally, the team is comprised of representatives from each of your departments, including legal, financial, procurement, human resources, etc. This team approach is essential because, one way or the other, everyone contributes to the safety culture of your company. In most cases, the contribution is positive, in others, perhaps not so much.

The team approach also assists with gaining broad insight of topics that need to be considered. Examples can include excessive absenteeism, high levels of attrition, product defects, customer complaints, accounting errors, procurement challenges, etc. As noted earlier in this book, while the overall theme of this book addresses safety culture concepts, these strategies can be applied throughout any process/program of a given company.

4.1: The Need for a Team Approach

This approach can also assist with long-term implementation of your safety culture improvement initiative after completion of the initial "heavy lift." I have found this provides incredible value to assist with sustaining traction, keeping individuals engaged, and maintaining momentum. Ideally, some members of the team are task level workers who can assist with providing positive peer pressure.

The team also needs to understand the importance of the effort, including viewing it as a project. By project, I am referring to development of a team charter, funding request, determining meeting schedules, establishing an overall timeline, and use of a Primavera P3 scheduling software (or equivalent) to track progress. Additional insight is provided later in this book regarding benefits from the use of these techniques.

As the owner, or equivalent, of the company, one of your most important roles early on in the process is to define the desired "end state." In other words, how will you declare success? This has proven to be a major stumbling block for many companies I have worked with. Hopefully, the following example illustrates this challenge.

Recently, I worked with the management team for a large company (over 1,500 employees) who had decided that they had a problem with workers not following procedures. This "conclusion" was based on reviews of a few incidents as well as feedback from representatives of their customer who periodically conducted oversight of work being performed. Even though there were thousands of work evolutions performed each month in a safe and compliant manner, senior management arbitrarily decided that procedure compliance had to be addressed company-wide. By the way, there was limited, if any, participation by task level personnel during the initial decision-making process.

Not to be deterred, management had posters developed, preached about procedure compliance during special all hands meetings, and issued numerous emails and memorandums on the subject. Taken at face value, one would think this approach should be successful. Unfortunately, the results proved to be the exact opposite. Due to not clearly bounding the problem nor involving personnel performing the work during initial planning stages for the campaign, there was no real buy-in by those routinely using the procedures in question.

Demonstrating improvements regarding procedure compliance also proved to be a logistical nightmare since the company, in essence, was attempting to prove a negative. In addition, since the company actually had a very high level of procedure compliance prior to rollout of the campaign, metrics established to monitor the effort could not demonstrate appreciable improvement.

Due to all of the energy expended on the campaign, including telling the customer how much better things would be in the future, this company was presented with an even larger problem. During numerous meetings I had with senior management, they expressed heartfelt concern as to how to respond to a future incident where procedures were not followed if they closed out the campaign. Namely, how could they ever declare success?

This gets even more interesting since it is not uncommon for personnel to not rigorously follow each and every procedure requirement every day. To be clear, I am not referring to willful noncompliance, which commonly isn't encountered in the workplace. What I'm addressing is the momentary mental lapse that can occur on occasion. In the majority of instances, these lapses do not result in an injury or major regulatory violation. When you compare the very real potential for a mental lapse to

4.1: The Need for a Team Approach

occur with the company's commitment to always comply with all procedural criteria all the time, the initiative was essentially doomed to failure.

In other words, they were in such a rush to roll out the campaign, they had invested little, if any, effort to clearly articulate what the problem actually was (versus what was perceived) as well as determining what success would look like (e.g., desired end state). This thorny issue eventually was overcome by events and slowly slipped off the radar screen. However, many of the individual employees still recall the overall process and continue to hold it up as an example of senior management not "getting it."

This example is reminiscent of the phrase: "Fire ... Ready ... Aim." If you're like the majority of readers, you're undoubtedly thinking ... shouldn't this read "Ready ... Aim ... Fire"? Congratulations, you're absolutely correct!

> ... use of a "Fire ... Ready ... Aim" strategy can result in a larger problem being generated than that initially perceived to exist.

From a military perspective, "Ready ... Aim ... Fire" refers to clearly understanding the objective, identifying the target, focusing on the target, and then actually hitting it. In the majority of instances, use of a "Fire ... Ready ... Aim" strategy can result in a larger problem being generated than that initially perceived to exist.

Without this foundational philosophy being applied, the company was stuck with attempting to demonstrate success where a true programmatic problem did not initially exist. I imagine even the legendary illusionist David Copperfield would have difficulty pulling that trick off. I am not attempting to make light of this situation. I am merely pointing out the incredible quagmire senior management can encounter if they launch a campaign, initiative, etc., without clearly identifying the challenge, defining the desired end state, and determining what is required to achieve same.

With all of this in mind, this needs to be one of your key roles throughout the development, deployment, and ultimate evolution of your safety culture initiative. Trust me, you do not want to be placed in the unenviable situation of having to explain to your board of managers, shareholders, etc., why you permitted a "Fire ... Ready ... Aim" approach to be utilized. As noted above, you need to ensure the end state is clearly defined, including the required actions to achieve the same. Although you have a series of roles to ensure the success of your company, I do not recommend that you attempt to run the team. I am offering this recommendation based on the following reasons.

— **Ownership**: Allowing the team to operate independently of your day-to-day "supervision" speaks volumes to the team members. This approach reinforces them being allowed to own the process as well as your confidence in their capabilities. This strategy commonly includes team representatives providing you with a periodic briefing regarding the team's process, near-term successes as well as speed bumps that are being encountered (more on that topic later in this chapter), and potential logistical challenges.

— **Capability to maintain a 50,000-foot perspective**: By avoiding getting caught up in the level of detail discussions and potential changes in team dynamics, it is much easier for you to maintain a big picture perspective. This also helps you be able to monitor progress of the overall goal(s). This is a critical role that needs to be retained at your level. This approach also allows you to provide a fresh set of eyes when being briefed by team representatives regarding their progress.

4.1: The Need for a Team Approach

— **Empowerment**: In addition to allowing the team to operate independently, ensure that you also empower them to make a series of decisions. I have assisted some companies with their safety culture journey where the president wanted to approve all aspects, including the agendas for team meetings. Having to be involved in every decision, and candidly poking your nose in where it really doesn't belong, can cause more harm than good. In extreme cases, this can result in the team throwing their hands up and just handing the project back to the president. In this scenario, everyone loses, including you and your senior team. This can also reduce the willingness of company personnel to volunteer for future initiatives.

— **Higher level commitments**: While the safety culture initiative is important, it really shouldn't be high on your radar screen of tasks that need to be routinely conducted at your level. After all, you are responsible for the overall performance of the organization. Examples include reporting to a board of managers/stakeholders, meeting with department heads, evaluating marketing presentations for potential new customers in key business sectors, monitoring financial projections, and the list goes on. Yep, you already have a very, very full plate. Let's not add to it.

Earn your leadership every day.

– Michael Jordan

4.2

Selection of "Champions"

Regardless of the industry or profession, there are always individuals who I refer to as the quiet leaders (e.g., champions). They are the people, commonly at the task level, whom others (e.g., peers) look up to and respect. To be clear, I am not talking about foremen, superintendents, or similar line management personnel. While there are numerous examples of these same managers also being champions, there are far more champions within a given organization who are not formally part of the line management chain.

I am a big proponent of continuous improvement, including my own abilities. To that end, I routinely attend and speak at safety conferences, serve as presenter for breakout sessions, and am an avid reader. I have also been involved in a wide variety of processes that utilize different techniques to identify personality characteristics in the workplace.

Examples include: Meyers-Briggs Type Indicator (MBTI), American Management Association (AMA) personality types, and DISC. The MBTI is a common assessment tool utilized by large companies, and is based on eight characteristics ranging from extroversion to introversion, thinking to feeling, etc. The AMA has a series of techniques to assist with

4.2: Selection of "Champions"

identification of personnel who like to lead versus those who are more comfortable being led. The DISC personality profile system consists of four main characteristics: dominant, cautious, inspiring, and supportive.

This is not an all-inclusive list, and there are numerous companies who offer personality assessment techniques. A common theme with any of the techniques is the benefits gained from being able to assess the different types of personalities within an organization. Such

> **Any company can benefit from the use of a technique to identify the personality types of individuals within their organization.**

an approach also assists employees with their career path as well as being able to recognize positions that may be a good fit for them, or areas that they want to improve upon. Any company can benefit from the use of a technique to identify the personality types of individuals within their organization.

By way of example, I'll never forget the experience of one of my good friends, John, who I have worked with for many years. He is very detail oriented, has a great sense of humor, and will gladly support any project. Having said that, John was not very comfortable having to speak in front of large groups. In other words, he was definitely more of a "behind the scenes" kind of individual. This is not to be viewed as a negative statement. It is simply an acknowledgement of his personality and comfort level.

A few years ago, his boss at that time decided her managers, including John, had to put on a play during the Christmas season, which included having to memorize a script and "perform" in front of approximately fifty people. Obviously, John wasn't very pleased with having to be part of this play and expressed his concerns with his boss. She responded by telling him to just suck it up and do it.

While John survived the process, he found the overall approach to be very upsetting. It also negatively impacted his routine work activities during that time period due to continually worrying about having to get on stage. More importantly, it generated a significant rift between John and his boss. Ironically, his boss was completely oblivious to how uncomfortable John was having to perform in the play. Candidly, if she had simply gotten to know John on a more personal level, she would have realized very quickly that being on stage was not a good task for him.

As noted earlier in this chapter, there are numerous techniques available to identify personality types. One publication I recall was written by Tim O'Leary and contains some memorable personality descriptors. The book is entitled *Warriors, Workers, Whiners, and Weasels, Understanding and Using the Four Personality Types to Your Advantage* (Tim O'Leary, 2006). Paraphrasing slightly, these personality types are described as follows:

Warrior: Warriors have self-confidence, a persistent strong sense of ability to succeed, and readily accept responsibly when something doesn't work out. Warriors never give up, never wear out, or take no for an answer.

Worker: Workers are the backbone of any company. They appreciate the importance of their work, take pride in their assignments, understand the importance of accepting responsibility for tasks not properly performed, and commonly have the best work/life balance of the four personality types. It is also not unusual for Workers to not express their concerns or feelings, expecting others to innately understand them.

Whiner: Whiners have the potential to be great workers, but their negativity and/or dissatisfaction commonly overshadow their performance. When compared to the Warrior or Worker, the Whiner will always look

4.2: Selection of "Champions"

for someone else to blame. This trait is especially apparent when they are at fault, which is often the case due to their negative mind set.

Weasel: Weasels are known for being extremely negative, and they operate from a profound sense of insecurity. In some instances, Weasels will act in an intentional manner to cause harm to the company or set their peers up for failure. In turn, their behavior can undermine the best efforts of a group or their peers.

During your safety culture journey, I encourage you to include as many of your Warriors and Workers (or similar personality types) as possible in the early planning stages. Speaking from a Warrior perspective (yes, I <u>definitely</u> am one), the majority of these individuals welcome this type of opportunity. They will feel challenged and look forward to assisting with development of the overall process.

In a similar fashion, the Worker provides consummate value by supporting the efforts of management and the Warriors with fleshing out the nuts and bolts of the initiative. These individuals can also assist with spreading the word throughout the organization, including sharing the overall concept, successes, and why they think the effort is important.

Ironically, I also encourage you to consider including Whiners in your efforts. There have been numerous instances when including this type of individual can eventually result in them becoming a contributing member of the team. In other words, they begin to adopt some of the desired attributes of a Worker. In that instance, you've not only reduced the number of Whiners in your company but also increased the number of Workers. I view this as a win-win situation for everyone involved, especially at the individual level.

Hopefully, you have few, if any, Weasels to contend with. Unless they choose to change their actions in the workplace, commonly the best option is to engage your Human Resources and/or legal organizations to determine the next steps. Contingent upon the individual, the final disposition may be termination. While this may sound harsh, the overall health of your company has to outweigh the needs of an individual who does not recognize the importance of being a value-added member of your workforce.

Train people well enough so that they can leave, treat them well enough so that they don't want to.

– Sir Richard Branson

4.3

ENCOUNTERING SPEED BUMPS DURING YOUR SAFETY CULTURE JOURNEY

While I'm highly confident that a series of the topics addressed in this book can prove to assist your efforts, I'm also equally confident that you are going to encounter a series of "speed bumps" along your safety culture journey. While I don't want to dissuade you from taking on this effort, I feel it is incumbent upon me to also point out that there may not be smooth sailing at all times.

> ... I'm also equally confident that you are going to encounter a series of "speed bumps" along your safety culture journey.

Candidly, I have found some authors or vendors of safety culture initiatives who are hesitant to address this topic at all. In other words, all you need to do is simply buy their product and everything will be fine. To be clear, there is a wide variety of techniques that may be able to assist with

4.3: Encountering Speed Bumps

enhancing safety culture programs. However, very few, if any, of them will work perfectly "off the shelf."

It is very common for an organization to encounter some level of pushback from their personnel when attempting to introduce a new process or expand an established process. This should not be surprising, since most individuals are somewhat hesitant to embrace change. One of the best examples is individuals who choose to be NIMBYs: Not in My Backyard. As noted in Chapter 2.3, they commonly will not want to support the "new" process. They may choose to quietly push back, remain disengaged, or downplay the effort behind their boss's back. In extreme cases, they can be very vocal regarding why the new process will not work, being a total waste of time, etc.

A slightly different personality type is what I refer to as WIIFM: "What's In It For Me?" While these individuals may not be at the NIMBY stage, they still need to be sold on how the new process will directly benefit them. In many cases, these individuals will not be very vocal regarding their concerns, but quietly choose to either not participate or only perform the absolute minimum.

So, how do you address these challenges? By way of example, consider the actions you may choose to take when encountering a speed bump while driving on a neighborhood street or in a parking lot. If you're like most of us, you will probably slow down and proceed cautiously until you cross over the speed bump. If you're familiar with the area, you might be able to plan your trip and use other streets, or portions of the parking lot, that do not have speed bumps. In other words, through careful planning, the speed bumps can be avoided completely.

You are also encouraged to consider practicing "deliberate speed." The concept of deliberate speed was recently illustrated by a group who was preparing for a firefighter competition. The competition consisted of ascending a large hill wearing their firefighting turnout gear while also carrying a variety of forestry firefighting equipment and then having to enter a helicopter positioned at the top of the hill.

As would commonly be expected, the team initially ran up the hill as quickly as possible. Due to their haste, team members began losing their footing, tripping, dropping tools, ultimately resulting in a very low overall score when compared to other teams. Naturally, they responded by attempting to run up the hill even faster during the next round of competition to make up for the "lost" time. This technique resulted in additional frustration, since more tools were dropped, with some team members actually rolling down the hill because they totally lost their footing. Naturally, their subsequent scores continued to decline.

After a series of attempts with no discernible improvement in performance, the team leader elected to have the team employ a technique he referred to as "deliberate speed." By moving slightly slower and taking their time, they were able to move more cohesively, keep their eyes on the path, and maintain their grip on the equipment. By use of "deliberate speed," the group ultimately received some of the higher scores in the competition. The team members subsequently shared this experience with peers at their "day jobs," with similar results being realized. The need to practice versions of deliberate speed when senior management is evaluating options when responding to challenges is a topic of many books addressing accident investigation strategies (Todd Conklin, 2012).

Another approach to smoothing out speed bumps is to hold a series of celebrations. Over the years, I have had the opportunity to work with a wide variety of craft personnel. Examples include pipefitters, carpenters, electricians, wiremen, sprinkler fitters, ironworkers, concrete masons, mechanics, janitors, and teamsters. Industrial settings have ranged

4.3: Encountering Speed Bumps

from commercial nuclear power plants, research facilities, meatpacking plants, chemical treatment plants, and large construction projects, to casino properties on the Las Vegas strip.

Regardless of the workplace environment, setting, or type of workers, everyone appreciates being offered the opportunity to join in a celebration. As an aside, a series of surveys have consistently indicated that most individuals really enjoy receiving apparel (e.g., ball caps, T-shirts, sweatshirts) as well as food! By the way, these celebrations are also another method to gain buy-in from the NIMBYs/WIIFMs personalities discussed earlier in this chapter.

The celebrations can address the accomplishment of a major milestone or something significantly smaller. In some instances, the celebration is accomplished through a companywide email, an article in the company newsletter, or a similar process. Whenever possible, I also encourage the celebration to be structured so that senior management has the opportunity to interact with task level/craft personnel.

One celebration approach that many companies utilize is holding a BBQ for their personnel. Whenever possible, it is ideal to have senior members of the team be the "servers" in the buffet line. One of my favorite memories addresses a retired Air Force general whose company was providing senior level support at a large facility conducting space exploration.

Whenever he was at the plant for this type of luncheon celebration, he could be found enthusiastically serving food to those in attendance while wearing a freshly starched chef's hat, no less. He always looked forward to these opportunities since it provided him with another method to chat with the task level personnel. Those being "served" were always very impressed that he was willing to wait on them. It was not uncommon for those in line to gently tease him by requesting more French fries, additional cheese for their hamburgers, etc.

As noted earlier, I've had the pleasure of working with a wide variety of companies across the United States for over thirty years, and I have participated in numerous celebrations of all shapes and sizes. One of the most memorable was an annual gathering of craft personnel, project management, and members of the senior management team. Total attendance commonly ranged from 150 to 200 individuals, including representatives from local law enforcement. The gathering celebrated overall project performance, including their ability to incorporate safety practices into routine work activities.

The event planning committee also invited a series of local vendors that carried safety products. Examples included numerous types of work gloves, fall protection harnesses, safety glasses, including prescription considerations, and various heights of stepladders. Since the company also had a comprehensive employee "wellness" program, attendees were offered the opportunity to check their blood pressure, review smoking cessation techniques, and consider options to help with weight loss. Local law enforcement representatives provided attendees the opportunity to navigate an obstacle course on adult size tricycles while wearing "drunken driver goggles." During this exercise, the police officers also discussed options to avoid operating a motor vehicle while potentially intoxicated, including use of a designated driver, or taxi cab service.

While the overall construct of the event readily embraced being able to inculcate safety into the celebration, this noteworthy accomplishment paled in comparison to their commitment to local charities supporting members of the military. Each year, the event planning committee partnered with two groups. The first group assists families at local military bases while the spouse is stationed overseas. The second group provides

4.3: Encountering Speed Bumps

service dogs for members of the military diagnosed with Post-Traumatic Stress Disorder (PTSD) or the equivalent. It was very common for a recipient and their service dog to attend the event, which further enhanced awareness of the importance of this charitable effort.

Now this is where the celebration got really interesting. Each year, the event committee also facilitated a "pie in the face" competition. The "pie" consisted of a traditional aluminum foil pie tin filled with a <u>lot</u> of whip cream. Attendees then had the opportunity to place bids, with the highest bidder having the opportunity to throw the pie into the face of the unlucky recipient. To help generate additional interest in the bidding process, numerous members of the management team would "volunteer" to have one of the pies thrown into their faces.

Needless to say, the bidding would gain significant momentum when the president and his senior team stepped onto the stage. To make it more interesting, bidders were permitted to have "group" bids, with the first series of bidders to meet a designated level having the privilege to throw the pies. Collectively, this portion of the event generated tens of thousands of dollars each year for their charity partners.

More importantly from a safety culture perspective, the pie in the face competition significantly strengthened the working relationship between craft and management personnel. After all, when the president of the company is willing to take a pie in the face from an ironworker, it doesn't get much better than that in my opinion.

This perspective was confirmed while I observed numerous instances of craft and management personnel interacting throughout the day, laughing about the experiences, with the mutual respect being readily evident. Another benefit was that this feeling of camaraderie did not stop at the end of the event but continued throughout the year.

I have also been part of more traditional luncheon celebrations that addressed achievement of milestones or similar accomplishments. The overall result was very similar to the experience described above. Namely, enhanced camaraderie, increased mutual respect, and recognition of the importance to function as a team.

In summary, the potential to encounter speed bumps can also be reduced by:

— Ensuring sufficient time is allocated for development of the strategy, including clearly defining the desired end state.

— Appreciating that some personnel may not want to initially participate or will only participate grudgingly.

— Establishment of a realistic schedule. I am always a fan of "under commit and over deliver." Another way to look at this is I would much rather be in the boardroom explaining how we were able to complete the task early versus having to address why original milestones were not met.

— Recognize that the NIMBY/WIIFM attitudes need to be addressed. However, you have to ensure that these individuals do not deter you from your ultimate safety culture goals.

— Celebration of near-term successes. While celebrating major milestones are always important, don't inadvertently overlook the opportunity to also acknowledge small achievements that are going to occur during your journey.

— Practice deliberate speed. In other words, do not be in such a rush to complete the initiatives that long-term sustainment is negatively impacted.

4.3: Encountering Speed Bumps

Consistent with the champions discussed earlier in this chapter, the potential for long-term success is increased by having individuals respected by their peers participate in the development and roll out of your safety culture efforts.

I have not failed. I've just found 10,000 ways that won't work.

– Thomas A. Edison

4.4

BEING "EMBEDDED" WITHIN A GROUP

In earlier chapters, I discussed the importance of senior management having the ability to connect with task level personnel. An additional option that has proven successful is to consider selecting an individual who has the support of the leadership team and place that individual within a given group for an extended period of time (e.g., 4–6 weeks).

> This technique can also greatly assist with improved working relationships between task level personnel and management, identification of process improvements, and enhanced safety.

While there a series of logistical challenges that need to be addressed, I can personally attest to the incredible insight that can be gained from this technique. This technique can also greatly assist with improved working relationships between task level personnel and management, identification of process improvements, and enhanced safety.

Very recently, I was asked to serve in this capacity for a large company. The balance of this subchapter addresses the overall process as well as the results of that effort.

4.4: Being "Embedded" Within a Group

— **Background**: The company had a group of approximately thirty personnel responsible for performing electrical work, including maintenance, repair, and installations of new equipment. Electrical voltage levels ranged from less than 50 Volts (V) to over 115,000 V. Work locations included conventional facilities, office space, warehouses, remote locations with extreme terrain, and elevations in excess of thirty feet in height. In other words, a wide variety of potential hazards and associated control techniques.

A few months before I was selected to be embedded with the group, they experienced a significant electrical near miss. Fortunately, no one was injured, but senior management established an additional series of interim controls for performance of electrical work. While the additional controls were successfully implemented, the senior management team was hesitant to subsequently "release" those controls without an independent perspective being provided prior to resuming routine work processes.

— **Approach**: After I was selected for this assignment, I held a series of meetings with the senior management team. I chose this technique so that I gained their insight regarding: 1) what they viewed as the potential problems, 2) what topics they wanted to be addressed, and 3) and how they would define "success." I intentionally utilized these meetings to provide my proposed approach and to ensure the resulting deliverables would meet their expectations.

Based on feedback gained from these meetings as well as my prior experience performing similar tasks, I developed a strategy whereby I would be present at all of the group's meetings. The first meeting began at 6:00 a.m. with the end of shift meeting being completed at approximately 6:30 p.m., including subsequent discussions I had with their line manager. I spent the balance of each day in the field with the crews while they were performing routine work activities. This included taking lunch breaks with them and sitting in on their safety meetings, planning sessions, and similar discussions.

At the end of each day, I generated an email that was routed to senior management. The daily message addressed work activities observed, feedback from crew members, as well as initial thoughts being generated. For ease of use, this was an ongoing message string, with the latest results located at the top. This method allowed recipients to readily review prior entries for additional perspective. To maintain transparency with the group, I intentionally included the line managers of the group that was being evaluated on distribution for the daily email.

During formation of the strategy, I also ensured that coverage could be provided for my routine day-to-day activities by other individuals. This proved to be instrumental so that I could solely focus on building an effective relationship with the group. Due to the remote locations for some of the work, I was able to use overnight accommodations in the immediate work area versus attempting to drive home each day. My ability to be one of the first persons to arrive each morning and one of the last to leave at the end of the shift proved invaluable to help establish a strong rapport with the group. In other words, I was making their workday a priority.

Towards the end of this assignment, I facilitated a meeting with all crew members in attendance. The purpose of the meeting was to gain their insight regarding why they should be allowed to return to their routine work practices. Other topics for the meeting included challenges beyond their control that were negatively impacting their ability to perform work as well as potential process improvements. This meeting provided a lot of insight regarding topics to be subsequently shared with senior management. Due to my extensive time in the field with the crews, they were much more willing to be open and candid with their thoughts

4.4: Being "Embedded" Within a Group

versus being with someone who had simply showed up for the meeting and asked what needed to be fixed.

All of this information supported development of a formal presentation to the company president and members of the senior team. I also ensured that the same presentation was subsequently shared with the group, including feedback from senior management.

— **Challenges**: While I was very enthused to perform this assignment, I was also cognizant of a series of challenges that I would encounter. First of all, the group had already been "reviewed" by other individuals, and they were feeling that all of their work activities were being second guessed. Additional challenges included: I was essentially an "unknown entity" to the majority of people in the group, they did not have a good working relationship with their senior management, and some individuals within the group were attempting to work around their line management.

Some of these challenges became self-evident when I first met with the group. After being introduced by one of the senior managers, I was immediately challenged by some of those in attendance. Questions included who I was, why I was there, and what my level of knowledge/expertise was with respect to the type of work they performed. By the way, this a great example of WIIFM discussed earlier in this chapter.

This type of response was somewhat expected since they were a tight-knit crew who took a lot of pride in their work, and I was definitely an "outsider." I responded to their series of questions by talking about why I was there (e.g., to gain insight on their great work practices that I could share with senior management) and also help them with rising visibility of problems/issues that needed to be addressed. In other words, I was forthright about not having expertise in their day-to-day work activities.

During the first few days, the reception I received from the crews while I was in the field was somewhat chilly, to say the least. However, during the second week, crew members became comfortable with my presence and began to offer comments regarding some of their work activities as well as solicit my input. Eventually I started to be viewed as one of the guys, including being gently harassed regarding my lack of a good haircut. Yep, I definitely valued this positive change in our working relationship.

— **Results**: The results of this approach provided a mechanism whereby a series of programmatic issues were identified and elevated to senior management. Referring to prior discussions in this book with respect to stressors, these programmatic issues were negatively impacting the workers, but they did not have the ability to resolve them. By the way, the group was thrilled with the results of my efforts. They also really appreciated being afforded the opportunity to participate in the meeting and addressing how to improve their processes, including potential options to address the stressors.

One area of particular note was when I provided a briefing to the group using the same presentation provided to the president. The group really appreciated this approach in addition to being viewed as part of the solution versus just being perceived as the problem. Subsequently, the senior management team actively worked with me and line management to formally release the additional controls that had been temporarily in place in addition to examining options to address the programmatic issues.

Perhaps of greater importance, this approach resulted in a series of programmatic issues being uncovered that would not have been identified during a routine audit or assessment. In other words, the results were definitely worth the effort.

4.4: Being "Embedded" Within a Group

While the accomplishments were noteworthy, this approach is not for the faint of heart. At the risk of appearing to beat my own drum, so to speak, any individual selected to perform this type of task needs to be comfortable working outside of their comfort zone. I am intentionally using that phrase since the individual will not, by definition, have a detailed understanding of the day-to-day work being performed. Another way to look at this is that you need to select someone who is willing to work without a net.

> Perhaps of greater importance, this approach resulted in a series of programmatic issues being uncovered that would not have been identified during a routine audit or assessment.

The next chapter addresses a series of considerations in support of your overall safety culture journey, including long-term implementation.

Fate whispers to the warrior ... you cannot withstand the storm ... the warrior whispers back ... I am the storm.

– Unknown

5.0

THE SAFETY CULTURE JOURNEY

As discussed earlier in this book, most companies expend significant resources conducting employee surveys, printing new safety posters, purchasing safety trinkets, and the list goes on and on. Many of these same companies then quickly declare success and are genuinely puzzled when long-lasting changes are not realized.

In some instances, this is a result of buying an "off the shelf" product that, by definition, does not address long-term sustainment. In most instances, you would be better off buying a dust mop since it can do a much better job of gathering dust.

> In most instances, you would be better off buying a dust mop since it can do a much better job of gathering dust.

I'm not belittling the importance of incentives, provided they are specifically tailored to your philosophy/culture. However, generic trinkets very rarely, if ever, provide long-term traction, versus those that are company specific.

Examples of value-added incentives can include stress balls with your company logo and small company "pledge bins" that can be affixed to company lanyards or apparel. While assisting one of a series of companies with their pursuit of OSHA VPP Star certification, I helped them design stress balls that were in the shape of a star (aligned with the VPP Star certification logo) and included their "motto" that addressed their safety culture. In addition to personnel actually using the stress balls, this item also served to help remind personnel of the company's commitment to safety.

Speaking of VPP, the process to successfully achieve certification commonly entails an 18-24 month effort. Having supported numerous companies pursuing this certification, one of the first areas of focus is working with senior management to gain their buy-in regarding the effort not being a thirty-day process. I have found this to be essential so that the management team does not inadvertently "push" for early completion.

Unfortunately, I have also worked with companies who chose to rush the process, wanting to complete it as soon as possible. This rush to complete the preparation efforts was readily identified by the independent review team during the subsequent formal review, ultimately resulting in the companies not being awarded the certification.

One of the most egregious examples I've encountered was a large construction company who was offered a financial incentive to submit their VPP Star application. However, the incentive was not tied to employee involvement or ability to actually receive the certification. For reasons that no one could fathom, two members of the senior team took it upon themselves to develop the application, with no participation by the workers. Unsurprisingly, the certifying authority rejected the application in its entirety and cancelled the pending onsite review. While the company got to retain the incentive, this approach generated significant friction with the workers, who were initially excited regarding the VPP Star certification process.

This strategy proved to be a significant hurdle when I was tasked with assisting the subsequent company with their VPP efforts. While the senior management had been changed out, the majority of the workforce remained in place. Needless to say, I encountered significant pushback, including some workers telling me they had no interest in supporting the process. This presented a very real predicament since the new management team really believed in the value of the VPP Star certification effort. This situation was compounded by the new management team committing to the certification effort with their parent organization. You could definitely say they were emotionally invested in the process.

Fortunately, I had a great team who was willing to meet with me and candidly share their concerns and beliefs. As mentioned previously, the beliefs and/or perceptions are very real to the individuals expressing them and must be treated as such if you want to shift their thinking. Thankfully, the majority of the team embraced safety and was willing to consider supporting the VPP effort. While they provided a conditional head nod, I also knew that more visible actions would have to be taken by the new management team to demonstrate to the workers that there would be a different approach than the one that had been taken in the past.

I was also fortunate to have the full support of the new management team, which proved essential for the next series of steps. To help team members buy into the VPP effort, I was able to obtain approval to send approximately twenty personnel to a national safety conference that also addressed the VPP certification process. This approval included funding to address labor costs, as well as travel expenses to fly the group across the United States. Prior to departing for the conference, I held a series of

5.0: THE SAFETY CULTURE JOURNEY

meeting with the team to address overall logistics, including who would attend different breakout presentations, what vendors to meet with, etc.

Perhaps the most intriguing part of this series of meetings was that the decision to pursue, or not pursue, the VPP Star certification effort was left up the group! That is correct. The new management team invested significant funds for this group to attend the conference, not knowing what the group's final recommendation would be. While this approach had inherent risk, it also had the benefit of clearly demonstrating to the group how important their support was. Fortunately, the group who attended the conference came back totally enthused and ultimately served in a champion capability during the certification efforts. By the way, not only did the company receive VPP Star certification on the first attempt, they successfully retained the same during a series of subsequent recertification reviews.

As part of my personal commitment to continuous improvement and learning, I sit in on numerous breakout sessions while attending safety conferences. On occasion, the breakout sessions are little more than a thinly disguised sales pitch. That is exactly what happened while I was at a safety conference a few years ago. I immediately left that session and happened to step across the hall to sit in on a presentation being led by a representative (she went by "Mattie") who supports the Mine Safety and Health Administration (MSHA).

That breakout session addressed the results of her research examining the safety culture of small "Mom and Pop" mines located in West Virginia. That phrase refers to mines that are operated by a small number of employees, with the husband being in the mine each day while the wife addresses bookkeeping, payroll, and other duties associated with managing the company. As is fairly common with this type of research, Mattie had a few graduate students on her team, and they traveled to numerous mines together.

While touring one of the mine properties, Mattie and her team were in the tunnel observing work practices. On more than one occasion, Mattie heard some of the miners call out "Sooie" or snort like a hog. The other miners would then laugh, cheer, or start snorting themselves. Since one of her graduate students was somewhat overweight, Mattie got very upset with these comments.

At the end of the shift, Mattie stormed into the owner's office and proceeded to read him the riot act regarding his workers treating her graduate student in such an unprofessional manner. The owner was very taken aback, telling Mattie that he had no idea what she was talking about. When Mattie shared the sounds and comments she heard in the mine, the owner started laughing. He then explained that the workers weren't mocking anyone. They are simply "coal hogs."

Since she hadn't heard this phrase before, Mattie asked him what he was referring to. "It is very simple," the owner replied. "Miners who are coal hogs work safe, keep an eye out for each other while on the job, and make quota." As an aside, when the miners make quota, additional pay is made available via a bonus. Now the concept of being a coal hog really peaked Mattie's interest, and she subsequently confirmed use of the same phrase at other mines.

Recognizing the incredible ownership being demonstrated at the worker level, she established a working group with representatives from a series of the small mines. The working group was comprised of primarily miners who worked in the tunnels each day. Working in conjunction with a graphics design firm, the group developed a decal, approximately 4" in diameter that could be placed on the side of a miner's hardhat. The decal had an illustration of the face and upper torso of a very muscular wart hog, complete with large tusks. The wart hog was wearing a tank top style white undershirt with a miner's hardhat on his head. On his very large left bicep, there was a prominent tattoo that read "Coal Hogs Work Safe."

As innovative as the decal was, the associated backstory proved even more interesting. The only way for a miner to obtain the coal hog decal was for the individual to sign a pledge. The pledge consisted of a few, very concise bullets that addressed workplace behaviors and attitude. To enhance support and buy-in, the pledge criteria were developed by the working group, not by the owners of the mines. The owners also agreed with the working group's requirement that only individuals actually working in the mines could be invited to take the coal hog pledge.

Mattie ended her presentation by sharing some successes she and her team had observed at mine properties that had implemented the coal hog pledge. During the question and answer portion, one of the attendees asked if she had any of the coal hog decals in her briefcase because he would like to have one. Based on the expensive suit he was wearing, he appeared to be a consultant or a business owner. She became very animated, stating that it appeared he hadn't paid much attention during her presentation. She then continued by asking the individual if he was a miner and if he had signed the coal hog pledge. When he answered no to both questions, Mattie looked him squarely in the eye and asked him why he thought he had earned the privilege to have the decal. Needless to say, the attendee really didn't have a response.

She then closed the session by noting she did not want to appear overly harsh, but the importance of this type of safety culture initiative cannot be inadvertently diluted by randomly handing out the decals. Such an approach could also be viewed as a personal affront to the miners who came up with the pledge. She received a standing ovation from all of those in attendance. I then had the opportunity to join her one-on-one for lunch.

Quoting the late Paul Harvey of talk radio fame, here is "*The rest of the story.*" During lunch, Mattie provided numerous examples of how miners were truly owning the process and were exceptionally proud to have the coal hog decal on the side of their hardhats. One of the examples that really resonated with me addressed new employees entering the workforce. Since the miners share breaks and lunchtime together, it was very common for current miners to approach the "new guy" within a couple of weeks of being hired and ask him why he didn't have a coal hog decal on his hardhat yet! The miners took this approach because they wanted to make sure everyone was working safe and watching out for each other. By the way, this is also a great example of positive peer pressure.

As we finished lunch, Mattie shared with me that the miners had voted to provide her with a coal hog decal. Somewhat puzzled, I smiled while inquiring how that was possible since she didn't work in a mine. She laughed out loud and then told me that the decal had been placed in a picture frame so that she couldn't put it on her MSHA hardhat. Even though Mattie had received many corporate awards for her efforts throughout her career, it was very obvious how much that framed coal hog decal meant to her. Later on in this book, you'll learn about how I received very unique recognition that I treasure as much as Mattie does her coal hog decal.

She also discussed her response to the individual who had asked for one of the coal hog decals at the completion of her presentation. Mattie was unabashed regarding her reprimand to the attendee since, from her perspective, he hadn't been paying that much attention during the session. More importantly, she refused to violate the agreement she had made with the miners regarding what was required to be issued a coal hog decal. This approach proved invaluable at a prior conference where some miners were in the audience. They really appreciated being able to observe her refusal to provide one of the decals to a company president who thought it would be "cool" to have one.

I subsequently shared this story with one of the companies I was assisting as one approach to engaging employees. You can imagine my dismay when I returned a few months later and realized that they had tried something very similar. However, they did not engage their employees during its development. They simply had copied a generic logo from a website versus one that would resonate with the workforce. Needless to say, the initiative fell flat on its face, and the management team was very frustrated. On the plus side, the company had a fairly small number of employees, and I was asked to facilitate an offsite meeting to examine different methods that could be implemented. With the help of the employees, the company ultimately came up with their own method that everyone could support. After approximately six months of "soak time," they began to realize results similar to those from the coal hog initiative at the coal mines.

By the way, it is not always necessary to have a formal pledge, or equivalent, campaign to enhance employee participation. In a lot of cases, employees will respond favorably with a simple request to give their word or share a handshake. I used this approach to address the change in senior management addressed in Chapter 2.0. In that instance, when employees spoke with me regarding their perceptions of Andrew, I casually requested them to give him a second chance. This included making an intentional decision to not overreact when Andrew did not initially respond in a manner that the employee expected.

When the employee(s) accepted this challenge, I also asked if we could shake hands on the agreement. In every instance, the employee readily shook my hand and also understood their obligation to the commitment. This strategy also supported the commitment by Andrew to be more receptive to techniques that enhance his interaction with the employees.

With all of this in mind, please appreciate the importance of recognizing your safety culture initiative as an ongoing journey. I also like to utilize the phase "persistent consistency," which was coined by one of my favorite authors, Zig Ziglar. In other words, remain resolute regarding your ongoing commitment to the effort.

This effort also needs periodic "pulsing" to ensure the topic remains fresh within your company. Without this perspective, it can get inadvertently overcome by events, project/scheduling pressures, and/or organizational changes. The balance of this chapter provides additional insight to assist with your efforts.

Discipline is choosing between what you want now and what you want most.

– Abraham Lincoln

5.1

NEXT STEPS

Due to the broad spectrum of topics that need to be addressed regarding next steps, there is not a single answer, not one solution, not a magic bullet. I freely admit that this statement is somewhat counter to the message shared by some safety program vendors. Namely, just buy this software, use a new checklist, etc., and everything will be fine. As discussed previously, there are a series of pitfalls attempting to implement a generic approach.

> Due to the broad spectrum of topics that need to be addressed regarding next steps, there is not a single answer, not one solution, not a magic bullet.

With this perspective in mind, the following steps can assist with the development of a safety culture enhancement strategy that is tailored to address the societal needs and organizational tempo of your company. For ease of use, they are presented in somewhat of a hierarchical structure.

— **Senior management support**. This commonly includes initial strategic planning sessions, discussions with the board of managers, etc., to conceptualize the overall effort. This step would also commonly address funding considerations.

5.1: Next Steps

— **Clearly articulate the desired end state.** The five safety culture implementation levels discussed in Chapter 1.3 can serve as a great basis to develop your vision. Your message can also reinforce some of the behaviors your company is <u>not</u> experiencing. Examples include not discounting the importance of safety, not endorsing short cuts to complete work, or not accepting minimal compliance. As a gentle reminder, you need to know where you're going so that you and your team will know when you get there.

— **Employee involvement.** Some of the best individuals to assist with your effort reside at the task level. They are the ones closest to the work and understand what really happens on a daily basis. As such, engaging them early on in the process provides significant value.

— **Establishment of a working group.** This approach assists with providing the broad perspective needed to ensure multiple viewpoints are being addressed. Including representatives from across your company is highly encouraged.

— **Conduct benchmarking visits.** Learning from other organizations can provide valuable insight regarding what may, and may not, work for your company. You are also encouraged to consider visiting companies that are in different business sectors.

By way of example, many years ago, the U.S. Army was examining methods to improve their ability to mobilize and demobilize their equipment. Rather than meet with other military branches, they chose to visit the Barnum & Bailey Circus. After all, this company had to erect tents, etc., in less than 24 hours and then leave the town by late Sunday night. Yep, they really understood how to mobilize. The techniques utilized by Barnum & Bailey proved very helpful to the members of the U.S. Army.

- **Development of a Primavera P3 (or equivalent) schedule**. Similar to any major project, use of a formal project scheduling software, including monitoring completion of major milestones, helps maintain emphasis and visibility on the effort. Completion status can also be routinely shared with participants to maintain awareness of key deliverables and status of the overall effort.

- **Marketing campaign**. There is a simple reason that major companies spend millions of dollars annually on marketing and advertising. It works. In a similar manner, having a series of methods to "advertise" your safety culture journey will greatly assist with enhancing employees' awareness and ongoing support.

- **Routine communications**. In addition to your marketing campaign, routine communications should also be considered. Examples include emails, articles in your company newsletter, frequently asked questions, and discussions at meetings. Collectively, these approaches can help keep the topic on the forefront.

- **Use of metrics**. Development of value-added performance measures (e.g., metrics) assists with demonstrating progress, as well as emphasizing benefits of the safety culture journey. Early identification of the appropriate suite of metrics can also assist with long-term sustainment. Additional insight regarding metrics is provided in Chapter 5.2.

- **Monitoring progress**. Using the major project analogy once again, it is essential that you and your team are able to routinely monitor progress for these efforts. This can be accomplished by ensuring a portion of your routine senior management meetings is reserved to address the current status of your safety culture journey.

5.1: Next Steps

— **Develop celebration strategies.** Celebration of milestones being achieved, goals being met, etc., are a natural extension of your marketing and communication efforts. While you obviously want to celebrate the completion of the overall safety culture initiative, it is also beneficial to celebrate small victories as well.

— **Establish techniques to support long-term sustainability.** Truly successful initiatives also address options to maintain momentum after completion of the initial initiative. Additional insight is provided in the balance of this chapter.

The suite of actions developed to begin your safety culture journey can also assist with establishing the overall framework and preliminary milestones. I highly recommend that this initial phase not be short-circuited or rushed to completion. In other words, remember to apply deliberate speed.

The task of leaders is to get their people from where they are to where they have not been.

– Henry Kissinger

5.2

METRICS

Perhaps one of the most significant safety culture topics that is routinely discussed, and rarely fully understood, is the use of metrics. Development of metrics has been addressed by numerous authors, including Dan Petersen. Some of his initial work is still viewed as foundational to understanding the concept of having metrics that provide value (Dan Petersen, 1996).

> Perhaps one of the most significant safety culture topics that is routinely discussed, and rarely fully understood, is the use of metrics.

While the overall concept sounds somewhat simplistic, there are a series of challenges that need to be considered. Metrics that are tailored to address company specific considerations provide considerable value, including helping monitor safety culture processes. While numerous companies offer software with the "perfect" metrics, this approach commonly turns out to be less than an ideal fit. The topic of metrics gets muddied even further due to companies having to monitor and report injuries and illnesses to the Occupational Safety and Health Administration (OSHA).

5.2: Metrics

The injury/illness information is compiled and submitted to OSHA through use of OSHA's Form 300, Log of Work-Related Injuries and Illnesses, commonly referred to as the OSHA 300 Log. The log summarizes injury/illness that occur throughout the year and is required by OSHA to be prominently posted in the workplace.

There are a series of additional reasons to monitor these statistics, including associated insurance costs, Experience Modification Rate (EMR) ratings, and ability to compete on certain scopes of work. While use of these OSHA statics in these instances does provide value by meeting a regulatory requirement, they do not portray the actual health of your company.

One of the most intriguing aspects of the OSHA statistics is the interdependency of the Days Away, Recordable, or Transferred (DART) rate and Total Recordable Cases (TRC) rate, based upon total hours worked. Namely, multiply the number of cases/incidents by 200,000 (an OSHA standard value) and then divide by the number of actual hours worked by the company. While this is the process required by OSHA for reporting purposes, it presents a series of challenges.

Due to the 200,000-hour value constant that is required to be utilized, smaller companies (e.g., those with fewer employees) will have higher values since they generate a lower number of work hours. To assist smaller companies, the Bureau of Labor Statistics (BLS) has developed information addressing statistical considerations for use by smaller companies. However, very few companies avail themselves of this information when calculating their rates. This appears to be due, in part, to not being a readily accepted practice across the majority of business sectors.

The same challenge occurs when comparing safety rates between different groups within a large company. By way of example, I worked with a company that monitored safety rates for each of its departments, with the OSHA statistics being reviewed monthly. At one of their senior management meetings, I was asked to lead a discussion regarding current performance, based on those same rates. While most of the departments had very low rates (less than 1.0), one department had a rate of approximately 43.5!

Understandably, the president became very concerned and demanded to know what was being done to address this "horrific" rate. The room then got very quiet when I explained that the rate was due to a single injury when an office worker experienced a small strain to their back while lifting a large binder off an overhead shelf. Due to the department in question only having a handful of people, the resulting rate calculation was significantly skewed when compared to the other groups within the company. One of the senior managers then commented that it didn't appear that the OSHA statistics were very helpful. I wholeheartedly agreed. I was able to use this discussion to successfully leverage a series of subsequent discussions regarding development of value-added metrics that could be monitored on a periodic basis (e.g., monthly, quarterly).

The topic of using OSHA statistics was also addressed during a breakout session at a national safety conference where I served as the presenter. Naturally, I began with the beginning, so to speak. This included addressing the overall history of OSHA statistics, how they are calculated, and linkage to insurance rates/industry averages, based on a given Standard Industry Code (SIC), etc.

Now this is where the presentation got really interesting. One of my slides summarized in table form DART rates for a series of fictitious companies (Company A, B, C). The example was based on a three-year period (i.e., Year #1, Year #2, Year #3). The table illustrated differing levels of performance for each company by year. Examples included the DART

5.2: METRICS

rate for Company A almost doubling over a three-year period, while Company C had a dramatic drop in their DART rate during Year #2. An example of the table utilized for that presentation is provided below:

	Three Year DART Rates for Companies A, B, & C		
	Company A	Company B	Company C
Year 1	4.6	2.5	4.5
Year 2	5.5	2.8	1.9
Year 3	7.9	3.5	3.2

I then asked attendees which company appeared to be improving as well as which company appeared to be declining. The majority of respondents readily concurred that Company A was not nearly as "safe" as Company C.

I then proceeded to explain that Company A actually experienced fewer injuries by Year #3. However, due to a major reduction in force during that same time period, the number of hours that were actually worked significantly dropped. As discussed earlier, when a company has a lower number of hours worked, their overall rates can increase, even though they had fewer injuries.

To reinforce the fallacy of being totally dependent upon OSHA DART/TRC rates, I then explained that Company C had more injuries during Year #2 but had doubled their workforce due to having to implement a second shift. I concluded this portion of my presentation by addressing the somewhat inverse relationship between work hours and rates. Candidly, at this point during my presentation, there was a growing sense of frustration by some attendees since I was definitely calling their view of OSHA statistics into question.

Finally, one the attendees raised his hand, asking what should be measured. I smiled, thanked him for the question, and then showed my next series of slides. The slides addressed a series of proposed topics, including having metrics that align with company core values, employee engagement, etc. After concluding my presentation, I was approached by numerous attendees regarding my availability to assist them with their efforts in addressing metrics.

Concerns regarding over reliance on OSHA rates have also been expressed by the Defense Nuclear Facilities Safety Board (DNFSB). The DNFSB, which is an independent oversight organization, was established by the U.S. Congress in 1988. This organization is tasked with providing advice and recommendations to the Secretary of Energy that addresses health and safety issues at defense nuclear facilities managed by the Department of Energy (DOE) and their contractors.

DNFSB Board Members are appointed by the President of the United States, confirmed by Congress, and are respected experts in the field of nuclear safety. The DNFSB is supported by a series of staffers, including those serving on detail assignments at select DOE facilities. In other words, the DNFSB carries considerable "clout" within the DOE and/or the National Nuclear Security Administration (NNSA).

One of the groups that the DNFSB routinely meets with is the Energy Facility Contractors Group (EFCOG). The EFCOG organization was formed by DOE Contractor representatives in 1991. The goal of the EFCOG is to enable sharing of DOE complex-wide information in support of excellence in DOE/NNSA activities and operations. The EFCOG includes a series of working groups and subgroups tasked with addressing numerous topical areas. Examples include nuclear safety, quality assurance, and worker safety.

5.2: METRICS

During the summer of 2010, the EFCOG Executive Council held their annual meeting in Washington, DC. At that meeting, the Chairman of the DNFSB presented a series of observations regarding the use of OSHA rates:

— The OSHA rates are not meaningful metrics of safety at defense nuclear facilities.

— There is concern by the board regarding an overreliance on OSHA rates leading to complacency and distracting from low-probability/high consequence events.

— DOE was encouraged to have their contractors focus on leading indicators addressing personnel, processes, and plants/equipment (Peter S. Winokur, 2010).

The DNFSB perspectives continue to assist the DOE and its contractors in their ongoing pursuit of value-added metrics.

While on the topic of metrics, I also want to share an experience I had while working with a consulting group based in Washington, DC. Among their deliverables was a Performance Indicator Report that was issued on a quarterly basis. The report addressed overall health and safety performance at numerous DOE/NNSA locations across the United States, and included a series of graphs portraying injuries, as well as DART and TRC rates. Due to the amount of information being evaluated, I would routinely lead conference calls with each of the sites to gain additional insight into what contributed to their overall rates, with particular focus on injuries.

One of the constants that I encountered when speaking with the site representatives was their ability to readily provide additional information regarding why their injury rates had increased. Examples that contributed to their increased injury rates commonly included use of temporary workers to address the increase in mission tempo, inclement weather conditions that weren't anticipated, introduction of new work processes that hadn't been appropriately evaluated, etc.

What really caught my attention was when I spoke with the same site representatives regarding a reduction in injuries. In other words, their injury rates had decreased. While this was a commendable outcome, what troubled me was that the site representatives usually had <u>no idea</u> what had contributed to the safer workplace. In the majority of instances, the site representatives were happy that their injury rates went down. However, they really didn't express much interest in learning what lead to the lower rates, and how to maintain the lower rates long term, including the development of additional metrics to monitor performance. When I pressed this topic further, some representatives responded that it would be "too hard" to look at other metrics. Other justifications included that they were already way too busy tracking completion of projects, lacked sufficient staff to monitor/update new metrics, etc.

Ironically, these same individuals passionately defended the need to monitor DART/TRC rates even though these metrics were actually providing little, if any, value regarding overall health of their safety programs. Yep, they were practicing the time-honored tradition of "this is how we have always done it." In other words, the sites did not recognize the importance of developing metrics that could be utilized to support enhanced safety and continuous improvement versus just counting injuries.

What is most troubling to me is that few companies have embraced the need to develop value-added metrics versus solely relying on OSHA statistics. Companies that have chosen to examine other strategies have commonly found that the benefits far outweigh the effort. To be clear, I am not suggesting that companies ignore the OSHA statistics. I am simply pointing out the significant challenges that can be encountered when a company chooses to only rely upon them to determine the overall health of their organization.

This challenge is further compounded by a lack of correlation between reduced DART/TRC rates and incidents that result in Serious Injuries or Fatalities (SIFs). Based on an BLS Report issued in December of 2017, the number of fatalities actually increased by approximately 7% during 2016, and the number of fatalities is at the highest since 2008 (Bureau of Labor Statistics, 2017). While there was a slight decline in reported fatalities during 2017 (5,147 versus 5,190 during 2016), the number of fatalities is still at a very high level (Bureau of Labor Statistics, 2018). However, during the same reporting periods, there was a decline in incident and injury rates.

This is puzzling for many companies since they expect SIFs to be reduced along with OSHA Recordable rates. However, this simply isn't the case, due to a series of contributing factors. Examples include stressors negatively impacting safety culture tenets not being consistently identified or effectively addressed, lack of a robust investigation process, and senior management not being sufficiently quizzical regarding underlying issues. The need to gain an understanding of these organizational stressors has been addressed by a series of authors, including Todd Conklin. In his book entitled *Pre-Accident Investigations, An Introduction to Organizational Safety*, he offers a series of topical considerations regarding human performance. Examples include personnel are fallible, attempting to place blame does not support understanding failures, and organizational systems/processes drive behaviors (Todd Conklin, 2012).

Perhaps one of the most glaring examples of this dichotomy is the Deepwater Horizon drilling platform disaster addressed in Chapter 1. On the day of the explosion, British Petroleum senior management representatives had flown out to the drilling platform to present a safety plaque. The presentation was for having operated for seven years without a lost-time injury. The company also monitored use of handrails while ascending/descending stairs, keeping lids on coffee cups, and prohibiting the use of cell phones while operating company vehicles. Based on these criteria, the company was performing work safely. Unfortunately, the Deepwater Horizon drilling platform disaster uncovered a series of programmatic level challenges that had not previously identified. Examples include ineffective procurement/receipt inspection process, emphasis on schedule, and personnel not feeling comfortable questioning some decisions.

When discussing this topic with senior management who emphasize OSHA DART/TRC Rates, I candidly nudge them, noting that it appeared they are potentially measuring the wrong "stuff." This inevitably leads to a series of subsequent discussions about what is truly important to their safety culture from a metrics perspective, including options to routinely monitor performance of those metrics.

Another approach I've taken with senior management is how safety can be viewed as supporting mission, positive outcome for performing work, etc. This commonly proves very insightful versus the traditional perspective of safety only being viewed as how many workers got hurt. Inevitably, the conversation assists with an enhanced perspective regarding safety aspects that should be considered during development of metrics.

As noted previously, use of value-added metrics can also assist with enhanced buy-in at the task level, maintain emphasis on your safety culture journey, and support long-term sustainment. To assist with continuous improvement goals, you are also encouraged to consider revising your metrics when the criteria are being met for an extended period of time. Examples of metrics that can be considered include:

5.2: METRICS

— Management presence in the field

— Participation by craft personnel during project planning

— Number/percent of work packages released to the field that were incomplete or inaccurate

— Closure status of employee suggestions and issues

— Percent participation for employee surveys

— Closure rate for corrective actions

— Tracking and trending of injuries and incidents

— Completion of period reviews for company procedures

Other topics for metrics that may prove beneficial are addressed in the balance of this chapter. Regardless of the method used to select your metrics, it is essential that they contribute to your goals, including monitoring the overall health of your company.

The goal is to turn data into information and information into insight.

– CARLY FIORINA

5.3

SAFETY CULTURE "THERMOSTAT" CRITERIA

Over the years, I have worked with a wide variety of personality types and management styles. While this book has intentionally been structured with a philosophical approach in mind for the big picture thinkers, I also appreciate there are a lot of linear thinkers out there as well. In recognition of the differing thought processes, the following safety culture thermometer criteria (e.g., checklist) are provided for your consideration. These criteria are not all inclusive and some may be a better fit than others for your company.

I am intentionally utilizing the phrase "thermostat" in recognition of encountering varying levels of implementation during your safety culture journey as well as the important role you serve in setting the bar regarding overall health, based on the criteria. As an aside, these criteria are not presented in a particular hierarchy or order of priority.

> I am intentionally utilizing the phrase "thermostat" in recognition of encountering varying levels of implementation during your safety culture journey ....

5.3: Safety Culture "Thermostat" Criteria

Use of these criteria can provide valued insight when evaluating the safety culture implementation levels addressed in Chapter 1.3. The conclusions gained from such an evaluation can also assist with providing objective evidence to support funding requests or discussions with your senior team addressing strengths and weaknesses.

I fully understand some readers (perhaps including you) will read the table of contents of this book and immediately jump to these criteria. Hopefully, those same readers will also take the time to actually read this book from cover to cover. Here's a hint: you want to practice deliberate speed when evaluating this checklist. Yep, that topic is also addressed in this book. I'll leave it up to you to go find it.

— Personnel throughout your company appreciate and support the need to be their brother's keeper.

— As the leader, you recognize the importance of employee involvement. In turn, your employees appreciate your willingness to support their involvement and respond accordingly.

— Safety initiatives are viewed as an investment, not a "tax." In other words, there is the understanding and commitment to ensure your safety organization and associated safety programs are funded appropriately.

— Small problems are identified and resolved before becoming big ones. This approach also reduces the potential for a significant adverse event to occur.

— People throughout your company are comfortable with raising their hands to discuss safety concerns. This is reflective of a learning environment as well as a Just Culture.

- The safety culture journey is clearly articulated so that everyone understands what is expected of them. The term "journey" is intentionally utilized to recognize that true success is an ongoing effort.
- Management commitment is readily evident throughout your company. This includes following through on employee concerns and proposed process improvements.
- Your employees recognize their duty and responsibility to pause work if it appears unsafe or requirements to perform the task are not clear. The employees feel comfortable doing this without fear of reprisal. This is another example of a learning environment as well as a Just Culture.
- The topic of safety is a standing agenda item for routine meetings. Safety topics are germane to the company, versus only being generic, and are flowed throughout your company.
- Safety remains one of the keystones of your company commitments. This includes not reducing commitment to safety when budget constraints are being addressed.
- Frequent and relevant communications are shared throughout your company. The communications address expectations from senior management, including being actionable at the employee level.
- There is a process in place whereby new employees are teamed with their peers in the workplace.
- A cohesive succession plan is in place for those approaching near-term retirement. This plan is routinely updated as required.
- The role of safety is owned by everyone in your company versus being viewed as solely the responsibility of your safety organization. In

5.3: Safety Culture "Thermostat" Criteria

other words, the safety organization is recognized as a resource, not a scapegoat.

— Craft personnel recognize and embrace the importance of their role in your safety program. They feel part of the overall process versus only being a worker on the shop floor.

— Personnel throughout your company routinely contribute articles to your company newsletter and other versions of company communications. Periodic articles are included addressing personal experiences that are heartfelt.

— Discussions at your senior management meetings address injuries and focus on prevention strategies versus just statistics.

— Your disciplinary action program is "just." In other words, the assigned discipline is consistently applied and is commensurate with the conditions surrounding the unsafe act.

— Feedback from external audit teams is embraced and incorporated to enhance overall performance.

— You and your management team spend value-added time on the "shop floor" to build camaraderie as well as gain an appreciation of how work is really done. In turn, your employees look forward to your visits versus impatiently waiting for you to leave their work area so they can relax.

— Personnel take pride in their work and their workplace. Examples include cleanliness of work areas and "tooling up" after the task is completed. Associated deliverables are well crafted and technically accurate.

— Recognition programs reward desired behaviors and are highly publicized. Awards are not directly contingent upon individual injuries or injury rates. In other words, employees who have been injured are not automatically excluded from potential recognition.

— Employee surveys are crafted and rolled out so that actionable information is compiled. The resulting information accurately portrays the perspective of participants.

— There is a safety suggestion program in place that is actively being used, provides value, and is supported by your employees.

— You ensure that process changes, once attained, remain embedded in your company. In other words, no thirty-day solutions or quick fixes that are not sustainable.

— Your vision of success, including your safety culture, is understood and embraced by everyone in your company.

You are encouraged to have your team review these criteria and revise as needed to better align with your company attributes or philosophy. I imagine you've also noticed that some of these criteria could be good candidates for company level metrics as well as for employee surveys.

However, to assist with subsequent evaluations, you are cautioned to <u>not revise</u> the individual criteria each year. This is critical so that confusion isn't inadvertently introduced when you compare results with data from previous year(s) as well as having your employees attempting to complete the associated survey. This approach can also inadvertently introduce bias when you are compiling results in support of tracking and trending efforts.

To enhance the overall effort, I have worked with a series of companies to develop tutorial guidance. The guidance consists of two to three paragraphs for each criterion as well as examples of successful implementation. This approach assists with conducting the evaluations as well as

5.3: Safety Culture "Thermostat" Criteria

enhancing overall consistently of evaluation results. At the risk of sounding repetitive, use of a team to develop the tutorial guidance definitely assists with evaluation criteria that are tailored to address the unique characteristics of your company.

You can never cross the ocean unless you have the courage to lose sight of the shore.

– Christopher Columbus

5.4

SAFETY CULTURE THERMOSTAT EXAMPLE

One of the benefits from a commitment to process improvement is the ability to measure current implementation status and then reevaluate the same on a periodic basis. To assist with prioritization of evaluation results, I recommend a color gradient process be utilized. Over the years, I have found a range from 3-6 color gradients is the most effective.

> One of the benefits from a commitment to process improvement is the ability to measure current implementation status and then reevaluate the same on a periodic basis.

While a larger number of color gradients can be utilized, experience has shown that increasing the number of the color gradients does not necessarily correlate to increased ability to evaluate results. In other words, the amount of effort required to differentiate between a broader spectrum of color gradients commonly is not justified by the end results. As an aside, the use of numerous color gradients can also prove somewhat problematic when presenting results via Power Point slides due to the limited ability to readily differentiate between subtle changes in colors.

5.4: Safety Culture Thermostat Example

In the majority of instances, these periodic evaluations are somewhat subjective in nature. This is intentional so that the participants conducting the evaluation have the opportunity to express how they "feel" regarding what was being observed or evaluated. When utilizing this technique, participants are commonly more self-critical when conducting the evaluations versus simply checking the box. This is yet another opportunity for you to engage and provide encouragement regarding the value provided from this type of evaluation. The following table has been developed to illustrate evaluation results for a fictitious company.

ID#	Example Criteria	R	Y	G
1	Personnel throughout the company appreciate, and support, the need to be their brother's keeper.		•	
2	Management commitment is readily evident throughout the company.		•	
3	The employees feel comfortable pausing work without fear of reprisal.	•		
4	There is a safety suggestion program in place that provides value.		•	
5	Process changes remain embedded in the company.			•
6	Personnel take pride in their work and their workplace.		•	
7	Feedback from external audit teams is embraced.		•	
8	The company disciplinary action program is "just."	•		
9	The topic of safety remains one of the keystones of the company.		•	
10	Small problems are identified and resolved before becoming big ones.			•
Red (R): significant implementation challenges, Yellow (Y): moderate implementation challenges, Green (G): effective implementation.					

A review of the evaluation results provides some rather interesting insights. Due to the level of camaraderie established at the task level, the company scored well regarding personnel watching out for each other. In other words, being their brother's keeper. The scores also reflect that the company appears to be effectively resolving problems and maintaining focus on long-term implementation of process improvements.

However, if I was assisting the management team at this company, I would be very concerned with personnel not taking pride in their work and why they feel the disciplinary action process is not being applied in a fair manner. More importantly, not feeling comfortable pausing work without fear of reprisal should be a major concern. In my years in the field, I have found these three topics to be codependent. When personnel do not feel that they have the support of their management, it is not uncommon for morale to decay to the extent that individuals do not care about their work. As previously discussed in Chapter 3, this can negatively impact financial performance due to rework, customer returns, and/or loss of customer base.

As the company began to successfully address the programmatic challenges associated with the evaluation criteria that had been rated Red, I would then encourage them to examine the criteria rated Yellow. By the way, I am not surprised that with the lack of visible management commitment, the topic of safety does not appear to be a priority.

I developed a similar strategy for a series of contractors required to evaluate a diverse suite of health and safety criteria on an annual basis, reviewing the collective results as a group. As the evaluations matured, additional criteria were incorporated that contained a series of more "soft" or "fuzzy" criteria that addressed safety culture. This was intentional so that participants could reflect on the actual health of implementation versus simply rendering yes or no decisions. This change initially proved somewhat challenging, but over the long-term, everyone agreed the expanded criteria provided enhanced value.

The capability to readily differentiate between effective implementation (e.g., color gradient of Green) and significant implementation challenges (e.g., color gradient of Red) also proved to greatly assist their efforts with respect to developing improvements at the programmatic level. Eventually, the annual evaluation process included the capability to "trend" results against previous years, which also proved beneficial.

Similar to previous discussions addressing the development of initiatives, you are encouraged to include your senior team when determining what criteria should be utilized. You are also encouraged to include task level personnel during the development process. Having task level personnel on the review team is always a plus for everyone involved. This approach commonly provides additional insight since peers are talking to peers.

Here's another hint. The topic of criteria for your safety culture thermostat should be included in your strategy during development and deployment of your safety culture initiative. This can also assist with addressing potential resistance from some members of your workforce.

You can be like a thermometer, just reflecting the world around you. Or you can be a thermostat, one of those people who sets the temperature.

— Cory Booker

5.5

Intellectual Curiosity

During my career, I've been fortunate enough to work with a series of visionary leaders. Some of the most memorable are those individuals conducting research at a series of national laboratories, including Dr. Glenn T. Seaborg, a world-renowned educator and research scientist. Recognition of his expertise included receiving more than 50 doctorates and honorary degrees. Of particular note is that a new element (Seaborguim) was named in his honor, the first element at that time to have been officially named for a living person.

He was also an advisor to ten U.S. presidents, including George H.W. Bush, and he traveled to over 60 countries promoting international cooperation for scientific endeavors as well as nuclear treaties. Among his many amazing traits was his ongoing fascination with research, an insatiable thirst for knowledge, and his ability to relate with individuals throughout the laboratory. I readily recall spending time with Glenn when he would patiently explain, using terms that even I could understand, some of his latest thoughts regarding his research.

In a similar manner, I have had the pleasure to work with a series of individuals with multiple PhDs as well as those who were recipients of

5.5: Intellectual Curiosity

the Nobel Peace Prize. Naturally there were varying characteristics with these individuals, including being analytically brilliant as well as having the ability to connect with graduate students. However, the one constant that I found to be universal was their intense intellectual curiosity. In other words, they were not comfortable with the status quo regarding conventional answers.

> However, the one constant that I found to be universal was their intense intellectual curiosity. In other words, they were not comfortable with the status quo ....

Ironically, I firmly believe all of us were born with this trait. As children, the majority of us were very comfortable exercising intellectual curiosity. Perhaps the best example is when a child asks "why" when encountering something new. While I don't have scientific data to support this, it also appears that some of us lose or choose not to utilize this skill as we get older. This could be a result of being picked on in school, discounted by a teacher, or simply a case of day-to-day life overshadowing the need to maintain intellectual curiosity. Or perhaps it is because we tend to become cynical as we get older, focusing more on seeing the world as it appears to currently exist versus how it could be. Candidly, the current tempo regarding instant media coverage with little, if any, associated detailed analysis can predispose the viewer to not consider how things could be.

To provide some insight, the following are some traits that I have found to be associated with those who embrace intellectual curiosity. I highly encourage you to acquire these individuals for your company whenever possible and nurture those with these traits that are already members of your organization.

— These individuals have the courage to challenge the status quo and intentionally push the barriers to stimulate thinking that results in new breakthroughs. They avoid "stove pipe" thinking and are not comfortable just gazing at their navels versus instituting action.

— When in a leadership role, which is quite common, these intellectually curious individuals focus on building high performing networks versus just a small team. They also look for methods to help build the careers of those around them. This includes looking across the company versus focusing solely on just their team members. In turn, this approach supports building depth on the bench and development of succession strategies.

— In spite of their high levels of intellect, these individuals inherently understand people and what motivates them. They commonly have a great deal of empathy and are able to help others perform at their best.

— By definition, intellectually curious individuals are risk-takers, promoting experimentation and testing boundaries without fear of failure. By the way, this skill is somewhat unique since fear of failure prevents the majority of personnel from taking action in the first place.

— As noted earlier, one of the challenges when interacting with these individuals is their propensity to refuse to accept the status quo. This includes not being content with the traditional attitude of "this is how we've always done it here."

— Intellectually curious individuals can be zealots regarding their opinions and engaging in philosophical discussions. However, they are also more than willing to share credit and readily welcome dissenting opinions.

5.5: Intellectual Curiosity

— Ironically, these individuals are also "doers" and very outcome driven. In other words, they are not willing to mainly function at the theoretical level.

From my perspective, intellectually curious individuals can readily compliment your safety culture journey, along with other significant endeavors. A word of caution. These individuals will readily challenge the "norm" regarding how things are done or what is being proposed. This should be viewed as an asset that can directly contribute to the overall success in a value-added manner.

***I have no special talent.
I am only passionately curious.***

– Albert Einstein

5.6

LONG-TERM CARE AND FEEDING

Before leaving this chapter, a word of caution regarding the importance of long-term care and feeding for your safety culture journey. As noted earlier in this book, there is the very real potential for emphasis to diminish after the initial campaign is completed. By way of example, I have worked with many companies addressing their VPP Star certification efforts, including techniques to sustain processes long term.

Unfortunately, I've also encountered some companies actually "losing" their VPP Star designation during subsequent recertification audits by independent auditors. I have personally witnessed the frustration, stress, and overall angst resulting from the loss of that noteworthy certification. In addition, this situation can quickly become a significant embarrassment for senior management.

In the majority of instances, the president is going to have to explain to the Board of Managers (BoM), or equivalent, and stakeholders what led to the loss of the certification. After all, the president had led a series of discussions with the BoM regarding the importance of the VPP Star

5.6: LONG-TERM CARE AND FEEDING

designation, including use of discretionary funding to support the effort. Then, the president had the pleasure of hosting an amazing celebration upon receipt of their VPP Star, which included posters, logos, and apparel with their VPP Star logo.

Over the next few weeks, there would commonly be numerous discussions about how the employees contributed to the success, and many of these same individuals would proudly share their experiences with their peers. In other words, everyone had caught the fever, and the enthusiasm was proving contagious across the company. This included new employees as they came on board.

Then ... slowly ... things began to change. With the subtlety of a slow-moving glacier, attitudes began to revert to those that were in place prior to beginning the VPP Star certification initiative. Some of the changes occurred somewhat behind the scenes, including having to reduce the budget for safety initiatives due to funding constraints. In this instance, the Finance Office may have made the decision without consulting the safety manager or the president's office. Commonly at this stage, the finger pointing begins as members of the management team attempt to maintain their position with the company.

> **With the subtlety of a slow-moving glacier, attitudes began to revert ...**

When viewed at the 50,000-foot level, there was a lack of overall situational awareness at the president's level regarding what was important, including initiatives that needed ongoing continuing emphasis. When this shift in attitude regarding the VPP Star certification initiative is brought to his/her attention, they are commonly shocked and dismayed to learn what has occurred. The next response is commonly what does it take to get the VPP Star processes back on track again, and they want to get the situation addressed immediately.

While the president is now fully engaged and wants near-term actions developed and implemented, ideally this situation can be avoided to begin with. This fictitious company will undoubtedly have a very different road ahead attempting to recover their certification, as well as reformulating their safety culture.

This scenario almost came to fruition for a company I was working with who had previously received VPP Star certification. In support of the upcoming three-year recertification process, I was tasked with facilitation of the overall on-site review by external auditors, including development of the entrance briefing for the audit team. As I began the effort, I came to the stunning realization that a series of processes supporting the company's VPP efforts had significantly degraded.

Examples included the company was no longer monitoring responses to improvement suggestions and attendance at safety conferences by company personnel had been significantly reduced. Perhaps the most troubling was that a highly touted safety council chaired by the senior team had not been convened in almost a year. When I presented my conclusions during a special meeting with the senior team addressing the upcoming VPP on-site review, there was concern expressed by those in attendance as to how this could have occurred. Fortunately, there was sufficient time to initiate a series of actions to address the challenges, including methods to ensure these topics remained on the forefront in outlying years. The heartfelt commitment by the senior team regarding course correction directly contributed to the successful retention of their VPP Star certification.

As an aside, when a company loses their VPP Star certification, the process to regain same will commonly take 12–18 months or more. As such, there is a significant incentive to retain the certification. It should

5.6: Long-Term Care and Feeding

not come as a surprise that the topic of VPP Star certification can readily be replaced with a series of other topics supporting your safety culture.

By way of example, I recently worked with a company that required a safety topic (e.g., safety share) be presented at each senior management meeting. The safety manager had dictated this approach be utilized but did not include other management team members in the decision-making process. He also did not offer strategies to implement the same. Unfortunately, this is another example of the "Fire, Ready, Aim" strategy discussed earlier in this book.

In numerous instances, the individual assigned to share the safety topic would simply "forget" to have one prepared. When actually presented, some of the so-called safety topics were not even applicable to the company. In other instances, the safety topic criteria that were presented actually violated company procedures. This situation became so inane that some administrative assistants would have safety shares that were generic and/or of questionable accuracy taped on the wall of their office so that their boss could quickly retrieve them to use at the meeting. This approach was made more egregious by the "presenter" of the so-called safety share having no additional insight regarding the topic being discussed. In other words, they simply read what was handed to them versus taking the time to research a safety share that was relevant to company processes.

Since the president had a lot of items that needed to be addressed during the meeting, no one was held accountable regarding an effective safety topic not being presented or not being accurate. In turn, this sent a very clear signal to the president's direct reports that the safety topic was no longer important. Not surprisingly, the concept of effective, value-added safety shares was not realized.

Other visible examples regarding the decay of this company's commitment to safety were a reduction of management presence in the field and lack of timely response to concerns raised by the employees. Again, these shifts are commonly not intentional. They are simply a reflection of external pressures (e.g., concerns from the BoM regarding profitability), organizational changes, as well as senior management being reassigned or leaving the company. In numerous instances, the company lacked appropriate performance measures (e.g., metrics) to monitor key processes and ensure that they continued to be successfully implemented.

Obviously, your company needs to develop an approach that is based on company norms and societal considerations. However, there are a series of themes that can provide value to assist with monitoring the ongoing health of your safety culture.

As noted earlier in this book, the topic of communication is paramount for any endeavor. Communication methods can include All Hands Meetings, Personnel Surveys, and periodic emails and/or articles in the company newsletter.

Another approach that should also be addressed is friendly "nudging" by the president to his/her direct reports, including how everyone is ensuring that safety remains on the forefront. A phrase that I have begun frequently applying is "intellectual curiosity," which was addressed earlier in this chapter. I also encourage factoring safety culture attributes into management bonuses when desired attributes are met.

Examples include exceeding the company goal for time spent in the field, successfully leading implementation of a new safety process, etc. However, management bonuses should never <u>ever</u> be solely linked to accident/injury rates. The reason I'm placing so much emphasis on this topic is that such an approach can result in the underreporting of

5.6: Long-Term Care and Feeding

incidents/injuries. This could be due to personnel not wanting to tell their boss they got hurt or the boss choosing to not flow the injury information "up line."

There are a series of topics addressed in this book that can assist you in your safety culture journey. I also wanted to offer the following considerations as you move from initial deployment to long-term sustainment. Additional insight is provided earlier in this chapter regarding the use of "thermostat" criteria to monitor the collective health your safety culture.

— **Metrics**: Over the years, there have been a series of quotes addressing some version of the phrase "what gets measured gets done." I freely admit that there is some validity to that concept, but it is equally important to ensure that the right topics are being addressed. In too many instances, companies simply grab some criteria from a random website to use as metrics. In my opinion, this topic is of such importance that it warrants a separate discussion that is addressed earlier in this chapter.

— **Connectivity**: While there are differing approaches to addressing this topic, the main theme is being able to establish a relationship with your employees. Techniques can include luncheons, time in the field, and spontaneous hallway conversations. Regardless of the mechanism utilized, the overall concept remains the same. Namely, having the ability whereby you are perceived as more of a peer by task level personnel versus only being viewed as the "big boss."

— **Commitment**: Referring to the "grit" discussion addressed in Chapter 2.0, you need to maintain ongoing emphasis regarding the importance of your safety culture journey. Candidly, this can prove to be one of the most significant challenges in your safety culture journey when having to balance other company priorities.

The following chapter contains a series of key thoughts that can support the beginning of your safety culture journey. They can also assist with long-term sustainment.

__Willpower by itself is not enough. If we want to achieve lasting change, we must have an effective strategy.__

– Tony Robbins

6.0

Not the End ... Rather the End of the Beginning

Hopefully, by now you have gained an appreciation of the importance of being able to establish an effective safety culture for your company, organization, or group. As previously discussed, the capability to successfully maintain a strong safety culture long-term is a commendable goal that is worthy of the pursuit.

> ... the capability to successfully maintain a strong safety culture long-term is a commendable goal that is worthy of the pursuit.

As you continue with your safety culture journey, be prepared to be pleasantly surprised when you encounter success where you least expect it. By way of example, I was recently working with a company that had a large craft contingent. One of the craft groups was responsible for performance of a wide variety of tasks, including carpentry, painting, welding, electrical, and maintenance of the air handling systems for a series of office buildings.

6.0: Not the End ... Rather the End of the Beginning

After completing the facilitation of a large meeting with the senior management team, I was walking down the hallway towards the exit. That is when I observed one of the craftsmen (Johnny) who was preparing to perform maintenance on an air conditioning system located above the ceiling. The activity required use of an eight-foot stepladder so that ceiling tiles could be removed and work on the overhead air conditioning unit performed.

Due to the diverse nature of their work, the craft had a series of procedures addressing safe work practices. Topics included selection of safety glasses and hearing protection, and proper use of hand tools. There were also procedures that addressed performing work at elevation, including use of scaffolding, extension ladders, and stepladders. Procedural criteria for use of stepladders included ensuring that the front and rear legs of the stepladder were fully opened, and the braces connecting the legs together were locked in place. This approach also ensured the stepladder remained stable during use as well as when personnel are ascending and descending the ladder.

When I approached the work area, I immediately realized that Johnny was using the stepladder inappropriately. He had left the front and rear legs of the stepladder closed together, with the back of the stepladder leaned against the wall, and he had ascended the first four steps of the ladder. I asked Johnny if he could climb down, and he got off the ladder. We then had a brief discussion regarding the safe use of that type of ladder. I had gotten to know Johnny fairly well due to being in a series of meetings with him and chatting together at different work locations. For additional insight on this technique, refer to earlier discussions addressing management presence in the workplace.

Johnny immediately recognized he wasn't using the stepladder correctly and appeared to be fairly embarrassed. He readily "owned" the error and thanked me for pointing it out. In turn, I thanked him for his attitude and his willingness to be open to the conversation. I then left the work area and continued with the balance of my day.

This company also had an expectation that any issue, no matter how small, including not following procedures, be immediately reported to the manager, foremen, etc. Johnny's inappropriate use of the stepladder readily met that expectation. Having said that, I chose to not formally report it since it was an unintentional error, and the situation was corrected on the spot. However, I did make a point to discuss the situation with the craft foremen and superintendents, presenting it as a positive example regarding being your brother's keeper. I intentionally did not mention Johnny's name since I wanted the discussion to focus on helping each other versus being perceived as ratting someone out. Recognizing the emphasis being placed by the company to follow procedures, I had previously cleared my proposed approach with the president.

Now, this is where the situation got very interesting. Unbeknownst to me, during the next routine safety meeting, Johnny went out of his way to discuss with fellow crew members his inappropriate use of the stepladder. He also made sure to compliment me for not blowing the event out of proportion and treating him with respect. I learned about this discussion when I met with his boss a few weeks later, who also thanked me for the way I handled the situation with Johnny.

This group also routinely contributes articles to the company newsletter addressing safe work practices. Imagine my pleasant surprise that there was an article, written by Johnny, discussing his incorrect use of the stepladder. In the article, he also thanked me for how I addressed the situation. Very pleasant surprise, indeed.

6.0: Not the End ... Rather the End of the Beginning

Their newsletter article also got the attention of the senior management team, who asked me to meet with them regarding how I could assist them with a similar approach. Ultimately, this turned into a very positive learning experience with senior management regarding how to interact with craft personnel as well as different options available to address situations encountered in the workplace.

Hopefully, you will hear of similar experiences within your company during your safety culture journey. When that occurs, and it should, don't forget to celebrate this type of learning moment. In this instance, the celebration could be as simple as recognizing it during a routine meeting. A little praise goes a long way, and individuals really enjoy receiving recognition for their efforts.

By now, I imagine you have a sense of my unwavering commitment regarding the importance of employees owning company programs. One of my favorite stories is when President John F. Kennedy was touring a series of National Aeronautics and Space Administration (NASA) facilities located in southern Florida. As part of his tour, he had the opportunity to speak with many individuals regarding the types of research being conducted. While at one of the maintenance buildings, the President observed a janitor mopping the floor. When asked by the President what his job was at NASA, the janitor set aside his mop for a moment. Looking the President squarely in the eye, he smiled and proudly replied, "I'm helping send a man to the moon." In my humble opinion, it simply doesn't get any better than that.

Like the majority of worthy endeavors, your ongoing safety culture journey requires commitment, tenacity, and heartfelt conviction. However, when viewed from the perspective of what can be gained for you, your company, as well as your employees, the results are more than worth the efforts. In conclusion, I want to leave you with some key thoughts, or takeaways, if you will:

— Determine why enhancing your safety culture is important to you and ultimately your company. Since the organization adapts to the persona of the president, it is critical everyone understands why this endeavor is worth the effort.

— Articulate your desired end state. Refer to the safety culture implementation levels discussed in Chapter 1.3 for additional insight regarding topics that can be addressed. Consider use of a small team to help "polish" your message as needed so that it will resonate throughout your company. This will also assist with increased buy-in, acceptance, and ownership.

— Identify champions to assist with the effort. In addition to members of your management team, actively seek out those who are respected by their peers and solicit their participation.

— Acknowledge that challenges (e.g., speed bumps) will be encountered during your journey. Regardless of the magnitude or number of these speed bumps, having a strategy developed in advance to address them will greatly aide in overcoming the same.

— Practice deliberate speed. In the majority of instances, more harm than good is generated when attempting to rush to completion. This commitment will also prove invaluable when examining options to address speed bumps, including the willingness of the team to discuss issues being encountered.

— Track the progress towards your goals and celebrate successes. While big accomplishments are obviously important, make sure to celebrate the small ones as well.

6.0: Not the End ... Rather the End of the Beginning

— Be prepared for the "long haul." The greatest value to be gained is what will be learned during your safety culture journey. In addition, truly successful safety culture processes include a continuous improvement component.

— Maintain your grit. It has served you well throughout your career. It will definitely prove invaluable for this endeavor as well.

Good luck with your efforts. You, your company, and your employees are worth it!

The following chapter provides some final thoughts. It also includes a discussion regarding how I acquired one of my most prized possessions.

The secret to success is good leadership, and good leadership is all about making the lives of your team members or workers better.

– Tony Dungy

7.0

Final Thoughts

As noted in the Introduction section, I chose to invest the time and energy required to author this type of book to offer insight into techniques that can assist you in your safety culture journey. To paraphrase some of the investors featured on the Shark Tank television series, *"you gotta have skin in the game."* Hopefully, I'm met that objective while also providing you with some points of view that may differ from those you may have encountered previously.

> To paraphrase some of the investors featured on the Shark Tank television series, *"you gotta have skin in the game."*

If some of my thoughts and/or suggestions do not initially appear to be a good fit for your company, you are encouraged to engage others and examine the topic further prior to writing it off completely. Conversely, I also want to caution you to avoid simply grabbing some of the topics in this book and directing your team to go implement them. In other words, don't overlook the importance of having solutions that are tailored to accommodate the specific demographics and societal considerations of your company. Since your safety culture journey is an ongoing process,

7.0: Final Thoughts

you will encounter the need to revise your strategy throughout the years to introduce new concepts as your safety culture processes continue to mature. When that occurs, I encourage you to review some of the other techniques provided in this book that weren't previously utilized.

I imagine you've also gained a sense regarding the vital importance of leadership commitment and employee involvement. When combined with a management team that embraces the need to have task level personnel join them at the table so to speak, you are well on your way. Functioning as a team is always the best approach, and your safety culture journey is no exception.

During my varied career, I have been honored and very humbled to receive numerous awards recognizing the support I have provided to a series of companies. Fairly recently, I was part of an awards ceremony where the president of the company presented me with a very unique piece of custom designed "desk art." This distinctive piece of flame shaped multi-colored lead glass is approximately two feet in height and makes a great impression on anyone who enters my office.

As I was completing this work assignment, the group also decided to hold a going away party for me at a local bar/restaurant. There were numerous people in attendance, including Scott, a retired member of the U.S. Navy. During his almost thirty years of dedicated military service to our great country, Scott was ultimately promoted to the position of Senior Chief. With his "seasoned" perspective, Scott was very selective as to who he chose to interact with at the management level. As such, I was very appreciative that Scott agreed to attend the going away party.

Towards the end of the evening, Scott asked if he could speak with me privately for a few minutes, and I naturally honored his request. Speaking very quietly, Scott thanked me for all of my support and guidance regarding his post-military career, and then he shook my hand. When I grasped his hand, I was stunned to realize that Scott had a "challenge coin" in his palm that he was presenting to me. For those readers who aren't acquainted with this practice, this coin is commonly reserved to recognize special achievement by members of the military, or in limited instances, private sector individuals. The presenter holds the challenge coin in their palm and exchanges it with the recipient via a hearty handshake.

I was obviously somewhat speechless to receive such unique recognition, especially when combined with Scott's propensity to speak his mind regarding management personnel who didn't "get it." After I recovered from my initial shock, Scott explained to me the history of the challenge coin, including making sure that I carry the coin with me at all times. This is important since if I encounter another individual with a similar challenge coin, I have to show my coin or be responsible for a round of drinks.

With all due respect to the president who presented me with the beautiful piece of desk art, I place a lot more value on the challenge coin I received from Scott. As noted throughout this book, I am unapologetic regarding my passion for personnel who are responsible for day-to-day performance of the work. As such, I carry that challenge coin with me wherever I go.

By the way, when I get the pleasure to meet you in the future, please ask me about the coin. I look forward to showing it to you. Yep, I think it is very cool.

In closing, I also wanted to touch upon the numerous quotes that are provided throughout this book. I have intentionally included them to provide perspective as well as to generate some additional food for thought. With that in mind, I want to leave you with one of my personal

7.0: Final Thoughts

favorites. It also provides some additional insight into my unbridled enthusiasm and personal commitment to assisting the safety culture journey of any company to the greatest extent possible.

You must get involved to have an impact.
No one is impressed with the won-lost record of the referee.

– John H. Holcomb

BIBLIOGRAPHY

Black, Brandon & Hughes, Shayne (2017). *Ego Free Leadership: Ending the Unconscious Habits that Hijack Your Business.* Austin, TX: Greenleaf Book Group Press.

Bolman, Lee G. & Deal, Terrence E. (2003). *Reframing Organizations, Artistry, Choice, and Leadership.* San Francisco, CA: Jossey-Bass.

British Petroleum (2010). *Deepwater Horizon Accident Investigation Report.* District of Columbia.

Bureau of Labor Statistics (2018). *Employer-Reported Workplace Injuries and Illness-2017.* Retrieved from https://www.bls.gov/news.release/pdf/osh.pdf.

Bureau of Labor Statistics (2017). *National Census of Fatal Occupational Injuries in 2016.* Retrieved from https://www.bls.gov/news.release/archives/cfoi_12192017.pdf.

Bureau of Labor Statistics (2018). *National Census of Fatal Occupational Injuries in 2017.* Retrieved from https://www.bls.gov/news.release/pdf/cfoi.pdf.

Collins, Jim (2001). *Good to Great, Why Some Companies Make the Leap ... and Other's Don't.* NY, NY: Harper Collins Publishers, Inc.

Conklin, Todd (2012). *Pre-Accident Investigations, An Introduction to Organizational Safety.* Boca Raton, FL: CRC Press.

Daniels, Aubrey C. (2016). *Bringing Out the Best in People: How to Apply the Astonishing Power of Positive Reinforcement.* (Rev. ed.) New York, NY: Mcgraw Hill.

7.0: Bibliography

Deepwater Horizon Study Group (2011). *Final Report on the Investigation of the Macondo Well Blowout.* Center for Catastrophic Risk Management, University of California, Berkley, CA.

Dekker, Sidney (2002). *The Field Guide to Human Error Investigations.* Burlington, VT: Ashgate Publishing Company.

Flight Safety Foundation (2005). *A Roadmap to a Just Culture: Enhancing the Safety Environment.* Flight Safety Digest, Vol, 24, No. 3, Alexandria, VA.

Gallup, Inc. (2016). *Employee Engagement in U.S. Stagnant in 2015.* Washington, DC.

Gallup, Inc. (2017). *State of the American Workplace.* Washington, DC.

Global Aviation Information Network, GAIN Working Group E (2004). *A Roadmap to a Just Culture: Enhancing the Safety Environment.* McClean, VA.

Goleman, Daniel, Boyatzis, Richard, & McKee, Annie (2013). *Primal Leadership, Unleashing the Power of Emotional Intelligence.* Boston, MA: Harvard Business School Publishing.

Institute of Nuclear Power Operations (2004). *Principles for a Strong Nuclear Safety Culture.* Atlanta, GA.

Institute of Nuclear Power Operations (2009). *Achieving Excellence in Performance Improvement, Leader and Individual Behaviors that Exemplify Problem Prevention, Detection, and Correction as a Shared Value and a Core Business Practice.* INPO 09-011. Atlanta, GA.

Institute of Nuclear Power Operations (2012). *Traits of a Healthy Nuclear Safety Culture.* INPO 12-012. Atlanta, GA.

International Atomic Energy Agency (1998). *Developing Safety Culture in Nuclear Activities, Practical Suggestions to Assist Progress.* Safety Reports Series No. 11. Vienna Austria.

International Atomic Energy Agency (1999). *Basic Safety Principles for Nuclear Power Plants, 75-INSAG-3, Rev.1, A Report by the International Nuclear Safety Advisory Group.* INSAG-12. Vienna, Austria.

International Atomic Energy Agency (2002). *Key Practical Issues in Strengthening Safety Culture: A report by the International Nuclear Safety Advisory Group*. INSAG-15. Vienna, Austria.

International Atomic Energy Agency (2002). *Safety Culture in Nuclear Installations: Guidance for Use in the Enhancement of Safety Culture*. IAEA-TECDOC-1329. Vienna, Austria.

International Atomic Energy Agency (2006). *Fundamental Safety Principles*. IAEA Safety Standards Series No. SF-1. Vienna, Austria.

International Nuclear Safety Advisory Group (1991). *Safety Culture: A report by the International Nuclear Safety Advisory Group*. Safety Series No. 75-INSAG-4. Vienna, Austria.

International Nuclear Safety Advisory Group (2005). *Safety Culture in the Maintenance of Nuclear Power Plants*. Safety Reports Series No. 42. Vienna, Austria.

Krause, Thomas R. & Bell, Kristen J. (2015). *7 Insights into Safety Leadership*. Ojai, CA: The Safety Leadership Institute.

Liberty Mutual Group (2018). *Liberty Mutual Workplace Safety Index*. Boston, MA.

Mathis, Terry L. & Galloway, Shawn M. (2013). *Steps to Safety Culture Excellence*. Hoboken, NJ: John Wiley & Sons, Inc.

National Aerospace Laboratory, NLR Air Transport Safety Institute (2011). *Just Culture and Human Factors Training in Ground Service Providers*. NLR-TR-2010-431. Amsterdam, The Netherlands.

National Commission on the BP Deepwater Horizon Oil Spill and Offshore Drilling (2011). *Deep Water the Gulf Oil Disaster and the Future of Offshore Drilling*. District of Columbia.

Petersen, Dan (1996). *Safety by Objectives, What Gets Measured and Rewarded Gets Done*, (Second Edition). NY, NY: International Thompson Publishing, Inc.

O'Leary, Tim (2006). *Warriors, Workers, Whiners & Weasels: Understanding and Using the Four Personality Types to Your Advantage*. Katonah, NY: Xephor Press.

Reason, James (2013). *A Life in Error, From Little Slips to Big Disasters*. Boca Raton, FL: CRC Press.

7.0: Bibliography

Reason, James (2016). *Managing the Risks of Organizational Accidents*. (Rev. ed.) New York, NY: Taylor & Francis.

Reuters World News (2016, December 8). *Japan Nearly Doubles Fukushima Disaster-Related Cost to $188 Billion*. Retrieved from https://www.reuters.com/article/us-tepco-fukushima-costs/japan-nearly-doubles-fukushima-disaster-related-cost-to-188-billion-idUSKBN13Y047.

Roughton, James and Crutchfield, Nathan (2014). *Safety Culture: An Innovative Leadership Approach*. Waltham, MA: Butterworth-Heinemann.

The Fukushima Nuclear Accident Independent Investigation Commission (2012). *The National Diet of Japan, the Official Report of the Fukushima Nuclear Accident Independent Investigation Commission*. Government of Japan.

United States Nuclear Regulatory Commission, Office of Nuclear Reactor Regulation, Office of Nuclear Material Safety and Safeguards (2005). *Guidance for Establishing and Maintaining a Safety Conscious Work Environment*. NRC Regulatory Issue Summary 2005-18. Washington, DC.

U.S. Chemical Safety and Hazard Investigation Board (2016). *Investigation Report Executive Summary, Drilling Rig Explosion and Fire at the Macondo Well*. CSB Report No. 2010-10-I-0S. Washington, DC.

U.S. Chemical Safety and Hazard Investigation Board (2016). *Williams Geismar Olefins Plant Reboiler Rupture and Fire, Geismar, LA*. CSB Report No. 2013-03-I-LA, *Case Study*. Washington, DC.

USA Today (2017, November 14). *NSTB: Amtrak's Systemic Safety Lapses, Lack of Equipment Caused Fatal Train Crash in Pa*. Retrieved from https://www.usatoday.com/story/news/2017/11/14/ntsb-amtraks-systemic-safety-lapses-lack-equipment-caused-fatal-crash-pa/862078001/.

Viera, Anthony J. & Kramer, Rob (2016). *Management and Leadership Skills for Medical Faculty A Practical Handbook*. New York, New York: Springer Science+Business Media, LLC.

Wilson, Larry & Higbee, Gary A. (2013). *Inside Out, Rethinking Traditional Safety Management Paradigms*. Electrolab, Limited: Belleville, ON.

Winokur, Peter S., PhD (2010). *DNFSB Perspectives on Metrics and Safety Reform*. Presented at EFCOG Annual Executive Council Meeting. Washington, DC.

Ziglar, Zig & Ziglar, Tom (2012). *Born to Win, Find Your Success Code*. Dallas, TX.